Introduction to the
Process of
Research:
Methodology
Considerations

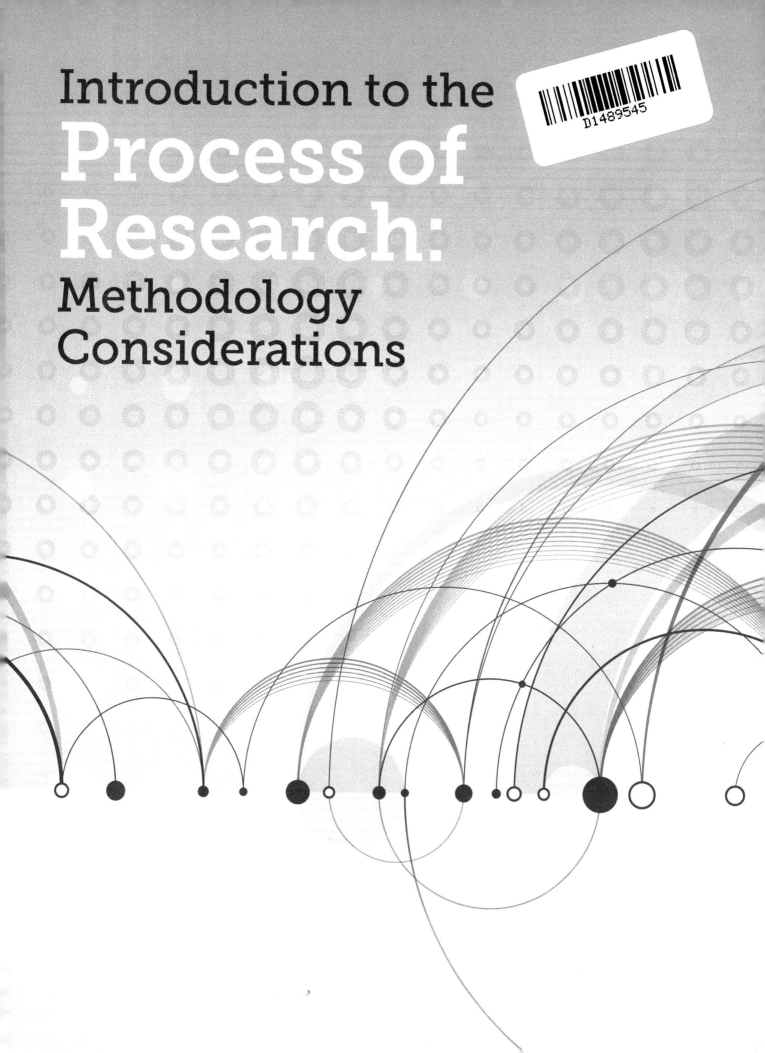

Introduction to the
Process of Research:
Methodology Considerations

P. Michael Politano,

Robert O. Walton,

Donna L. Roberts

Design and editing: Gratzer Graphics LLC
Editing: Gratzer Graphics LLC and Ruth E. Thaler-Carter

Published by:
Hang Time Publishing, Ltd. Co.
c/o P.M. Politano
171 Moultrie St.
MSC 24 Citadel
Charleston, SC 29409

Politano, P. M., Walton, R. O., & Roberts, D. L. (2017). *Introduction to the process of research: Methodology considerations*. Charleston, SC: Hang Time Publishing.

To correspond with the authors, please send email to: politanom@citadel.edu or waltonr@erau.edu.
Printed ISBN 978-1-387-10057-6
Ebook ISBN 978-1-387-10058-3

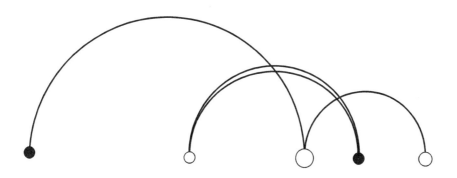

Dedication

To my wife Deb and my children: Holt, Amani, and Gwynn.

—P. Michael Politano

To Mom and Dad, my wife Petra and children Kinga and Reese.

—Robert O. Walton

To Sergio, forever the wind beneath my wings.

—Donna L. Roberts

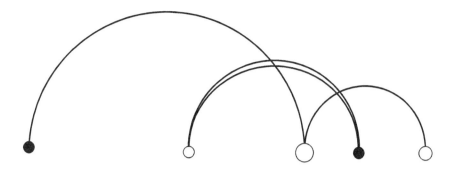

Contents

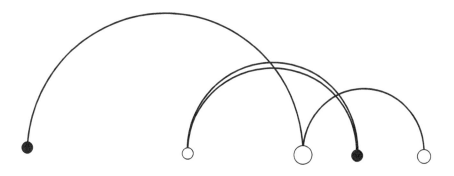

Foreword

Students always seem to struggle with how to conduct research. From the first moment that a freshman in college steps into the classroom (and even earlier now, in accelerated high school programs), he or she is asked to write a research paper on this or that subject. They write many of these papers in the form of a simple book report. Even though the students think that they have written research papers, they have not. While they may have searched an online library for reference material, the simple regurgitation of this information does not constitute research. Research is the process of discovery of new knowledge, either by synthesizing what has already been published or by collecting data to confirm or extend what is already known about a subject.

We hope you find this book on research methodology to be an easy-to-follow guide to the research process. We have tried to present the basic research process in a somewhat informal format that is intended to be easy to read and comprehend.

There are several important keys to this book. Flow charts in Chapters 1, 5, and 6 visually portray the research process relative to quantitative and qualitative methods. At the beginning of each chapter, you also will find a flowchart of the overall process of research, with specific sections of the flowchart highlighted to match the content of that chapter. This should help keep you on track and reduce confusion as you begin the journey of understanding research.

We intentionally did not include statistics in this book (but did reference various methods), assuming that you have a basic understanding of the common statistical methods and how to perform statistical analysis.

Research is an undertaking that values obsessive-compulsive behaviors. As such, it follows a fairly rigid set of steps from beginning to end. You will find those steps detailed in these chapters. Once you understand those steps and their place in the overall process, you will be a budding researcher yourself.

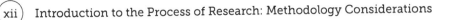

Throughout the book, we use various published studies to highlight and support the subject that is being discussed. While we do not go into a great deal of detail on each study, we encourage you to search for and read the studies in their original journal presentations. Many are classics, and all are quite interesting (citations are provided in the reference section at the end of the book).

—*The authors*

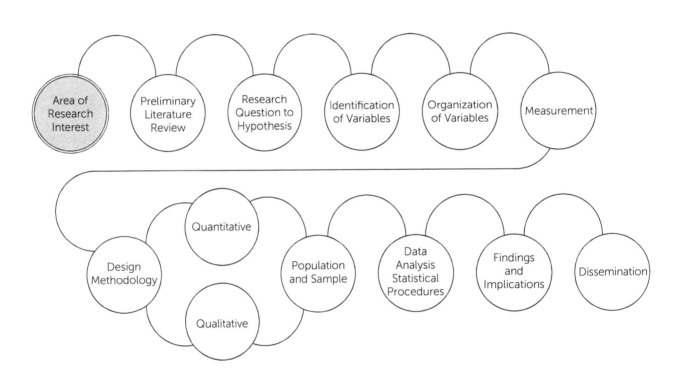

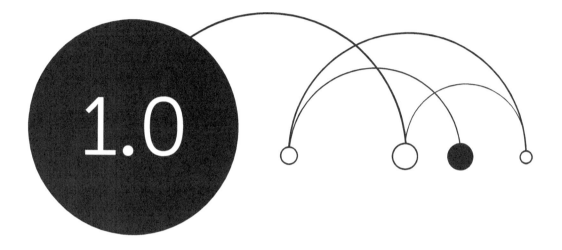

Research Fundamentals

If there is one trait that best defines a researcher, it is the ability to concentrate on one subject to the complete exclusion of everything else in the environment. This sometimes causes researchers to be pronounced dead prematurely. Some funeral homes in high-tech areas have started checking résumés before processing the bodies. Anybody with a B.Sc. or experience in computer programming is propped up in a lounge for a few days, just to see if he or she snaps out of it.

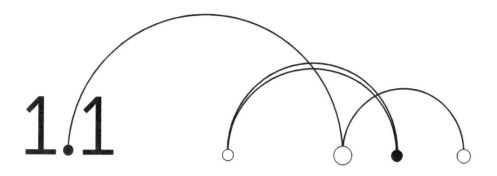

Why do research?
What is gained?

We would bet—if you are like so many of the students we have taught over the years—that you are taking a course in research methods (and, probably earlier, a course in statistics) because it is a required course. Were it not a required course, you would probably avoid it like the bubonic plague. As one of our students remarked in class a few years ago, "Taking stats and research methods courses is somewhat further down on my list of 'favorite things to do' than a root canal."

If this is not a required course and you are taking it voluntarily, then you have probably been accused by your peers of having a screw loose and/or probably qualify for being a "nerd." After all, nobody in their right mind would take a research methods course of their own free will, right?

It is admittedly a difficult subject, more so because it is often viewed by students as a hurdle to get over, a necessary evil, a boring course, something they will never use in the real world, information that will not impress a potential boy/girlfriend, etc. We hope this is not your perspective on starting out. If it is, or this is even close, we hope we can point your perspective in a more positive direction. Indeed, we hope that by the end of this chapter, certainly by the end of the book, you will view the pursuit of research (and statistics as a tool in that pursuit) as one of humanity's most noble endeavors.

Let's look at some of the reasons why human beings engage in research.

Could a Chimpanzee Compute E = mc²?

As human beings, we engage in research because we can. We have the intelligence and have developed the tools to ask questions and seek answers in reasonably objective and unbiased ways. Research helps us develop tools to make our lives better and our world a safer place to live.

To the best of our knowledge, humans are the only species to engage in research. We say "to the best of our knowledge" because someday we could be surprised to find, for instance, that dolphins have been studying us the whole time we thought we were studying them.

When we look at other species, we certainly find that some engage in experimentation. Chimpanzees' use of carefully selected sticks to dig out termites and sea otters' use of smooth rocks to bang open oyster shells are examples of animal behaviors born out of experimentation. Moreover, they even teach these behaviors to their young. However, is this research?

Such behaviors would constitute **research** if they were undertaken in a systematic way and directed toward establishing facts or principles related to a general field of knowledge (Webster, 1980). We think, however, that a chimpanzee's use of a stick to entice termites represents a more happenstance acquisition of a useful behavior aimed at filling its tummy, rather than a systematic investigation into factors such as optimal stick attributes, optimal patterns of stick-wiggling producing the greatest attraction to termites, the nutritional value of termites versus the effort required to attract them with a stick (calorie out-calorie in expenditures), or where termites fit in the overall food chain.

Barring some fantastic discovery of dolphin research files on humans at the bottom of the Great Atlantic Rift, we can safely say that we are the only species that conducts research in the fullest sense. We do it because we can.

The next question to ask is, "Why?"

"Mommy, Mommy. What Is the Moon?"

This very question contains part of the answer—"What is …?" How many times have you heard parents complain about the incessant why's, what's, and how's from their 2- or 3-year-olds? We appear to be born with an insatiable curiosity to know the why's, what's, and how's of things: Why do it this way instead of that way? What is the relationship between this and that? How did that happen? How do I compare to other college students in my attitudes about grades, dating, politics, and so on?

And what a range of questions we can ask—from the absurd ("Why are we told to quit acting smart in school? Should we act dumb, instead?"—from Gallagher) to the more abstract ("What is the origin of the universe?") to the applied ("How can we reduce carbon monoxide emissions in cars?") to the commonplace ("Which do you think is better—Coors or Budweiser?"). Our why-what-how questions may become less frequent as we age, but those questions often become more focused, intense, or worldly ("How do we cure AIDS?")

This insatiable curiosity drives a good deal of our research, but it is not the only motivating factor. In addition to being curious, we are very self-protective as a species. We have to be! We lack a stellar array of physical equipment—in the form of claws, teeth, and speed—that we could use to survive in a hostile environment in direct physical competition with other animals, particularly the meat-eating kind.

This strong inclination toward self-protection, given our physical vulnerability, has also served us well. We, as a species, have spread far and wide into a vast range of physical and ecological niches, from the frozen arctic tundra to the Sahara sands. While we have developed the skills and means of dealing with things such as inclement weather conditions or large predators, our greatest battles have been against the ever-present and often-deadly microorganisms that thrive around us and in us. Think of the bubonic plague in Europe from about 1346 to 1353, when upward of 60% of the population of some town was wiped out (Benedictow, 2005). Imagine if tomorrow, 60% of the people where you live just disappeared!

Take a moment, adopt an evolutionary perspective, and imagine what it would have been like to have lived 3 to 4 million years ago, given the conditions of the time, and how remarkable it is that small humanoids like our ancestors managed to survive at all, much less rise to such prominence. If 100 of us were suddenly plopped back in time to that period, even armed with high-powered rifles, could we survive? We would give such a group less than a 50-50 chance (probability in action) of making it, particularly after the ammunition runs out.

Stop and think for a moment about the number of self-protective mechanisms that you come into contact with on a frequent, if not daily, basis that have resulted from research efforts. Just to list a few: grounded electrical circuits, the seat belts and air bags in your car, the inoculations you carry around in your body, the ability to fly, the lead apron you wore at the dentist for your last X-rays, the X-rays themselves, the government tests performed on the meat that made up your McDonald's hamburger lunch, nearly everything in your medicine cabinet, the filtered and treated water coming out of your faucet … the list goes on and on. We are so very protective as a species.

Polio was the scourge of the world when one of your authors was growing up. During outbreaks, church services were canceled, swimming pools closed, movie theaters shuttered. Nobody knew what caused polio, so the idea was to keep people apart to prevent its spread.

The father of one of your authors was a physician and looked after polio patients at a hospital for crippled children near the town where he practiced. The hospital was filled with row after row of Emerson-model iron lungs, their mechanical hiss providing the necessary vacuum and alternating pressure needed to keep lungs functioning. Children in these mechanical devices saw their world by looking in a mirror over their heads while lying flat on their backs, 24/7. Other children—luckier children—could get about in wheelchairs with portable iron lungs weighing down their chests, or in leg braces and crutches (think President Franklin D. Roosevelt).

Dianne O'Dell contracted polio at age 3. She spent the next 60 years of her life lying flat on her back in an iron lung. She died when an electrical storm knocked out the power to the iron lung—she suffocated. People such as Dianne O'Dell were common images in the days of polio; images now nonexistent. The cure was born of necessity (self-protection) and curiosity (what causes this disease?).

All of those vaccinations you have had were discovered through the same process: self-protection and curiosity. When you stop and think about it, nearly everything you touch, eat, buy, sell, or throw away in the trash on a daily basis that is not directly of the Earth is probably the result of somebody's research.

Are you wearing your hiking boots today? Look at the bottom and see if the yellow Vibram label is embedded in the rubber. In 1935, a team climbing Mt. Rasica in the Italian Alps was caught by storms. Six climbers fatally succumbed to exposure and frostbite. They were wearing the standard leather-sole hobnail boot of the days. It was a tragedy that shook the climbing world.

Vitale Bramani, an Italian alpinist and friend of the six climbers who died, set out to build a better mousetrap—that is, to design a shoe tread and rubber compound that would give better grip, more flexibility, and more adhesion to ice in cold weather than what existed in standard shoes. In effect, he hoped to create a shoe that would give climbers a greater ability to climb up or down in extreme weather conditions and avoid such a disaster again.

Working with the tire company Pirelli, he developed the waffle-stomper design of deep, self-cleaning lug soles. Years ago, Bramani started what you are wearing today (ViBram) that helps keep you from slipping on wet, muddy, or icy ground in response to the question, "How can we make climbers safer?" (personal communication, Maura Folan, *The Mountaineers*, 1999).

Has Monday Night Football Started Yet?

This is a rather sexist introduction to this section, given that it is usually men who achieve a new level of meaningful existence during football season. The point, however, is not the meaningfulness of male existence, but rather the extent to which we—regardless of gender—look forward to familiar activities that are comfortable, entertaining, and enjoyable.

We are, by and large, comfort-bound creatures in westernized societies. We love our luxurious cars, our comfortable chairs, our warm houses on a cold night, our various and sundry entertainments, fast (and convenient) foods, etc. Where would you be without your down-filled parka on a cold day, your air conditioner on a hot day, your football games or soaps on a lazy day, your CDs as you read this text?

A good friend of one of your authors goes camping—at a campground where they set up the tent for you and the tent has electricity, hot and cold running water, and color TV. That is his idea of "roughing it" and he loves it—he would not even consider a *real* camping trip! Our wonderful comforts are, in nearly all cases, the results of entrepreneurship turning basic and applied research into marketable commodities. Enterprising individuals have translated much of the research findings related to curiosity and self-protection into best-selling comforts and, as a consuming public, we love those translations.

Does this mean that we should add greed as one of the motivators for research? While making money may seem a less-than-noble reason for conducting or applying research, it clearly has been a driving force behind much of what we enjoy, or even need, today. Drug companies, looking for new vaccines for AIDS, SARS, cancer, etc., certainly plan to make a profit from such discoveries, apart from any altruistic motivations, so, yes, greed is a driving force behind the advancement of research—a consuming public's greed for new toys and corporate greed for a better bottom line. But let's not call it greed. Let's call it entrepreneurial enterprise—sounds more noble!

Short Twitch—Long Twitch: Genetic?

It was "Jimmy the Greek" who made some observations, back in January 1988, about racial differences that favored one race over another in football. What he said is not worth repeating (you can look it up online). Nevertheless, it enraged a number of people and led to the end of his career as a CBS football analyst.

Despite what happened to "The Greek," the notion of superior short-twitch muscles in African Americans—thus favoring them in football and short sprints, but not swimming or long-distance running—still holds sway as a racial stereotype. This stereotype seems to persist despite the dominance of Kenyans in long-distance events—a long-twitch–advantaged sport—and the recent Olympic gold medal (Beijing Olympics) earned by the American swimming relay team that included an African American athlete (Cullen Jones). Clearly, there are no hard-and-fast stereotypes.

A popular book said that men are from Mars and women from Venus. Folklore has it that blondes have more fun and are not as smart as females with other hair colors. The argument of racial differences in the distribution of intelligence seems to resurface periodically, fanning the nature-nurture debate.

Because we tend to "flock together" with others having similar beliefs, values, and behaviors, we also tend to see other "flocks" as fundamentally different from our flock, whether those differences are valid or not.

In the early part of the 20th century, ethnic slumming—i.e., strolling through ethnic neighborhoods—was a frequent leisure-time pursuit for the distinctly "American" (Cocks, 2001). One of your authors is often amused that the two favorite neighborhoods to visit, because of cultural distinctiveness, were Italian and Asian. Having grown up Italian and American, he is not sure that he appreciates just how different "his people" were culturally. Nor does he see a culture that gave rise to the likes of Michelangelo, Da Vinci, Galileo Galilei, etc., as intellectually inferior, as would be suggested by immigration quotas from the early part of the 20th century that arose, in large part, from the eugenics movement.

These observations are all part of the stereotypes many of us hold toward other groups of individuals. One function of research should be to clarify the existence or nonexistence of differences, thus elucidating the playing field for ourselves and for all others.

How many times have you been involved in a conversation about some controversial topic, such as the relative strengths and weaknesses of the genders? Did the conversation get somewhat heated? Did somebody mention, at some time, a gender difference or a strength or weakness that raised your blood pressure just a little? Did somebody mention a supposed strength/weakness that he or she then attributed to some unnamed research, as if to give the strength/weakness validity beyond that person's stereotypic bias?

It seems that whenever we want to make a point that reinforces our own peculiar biases, we have a tendency to say, "Well, research has shown …" This recourse to some unnamed research as a final authority for individual or group beliefs seems part and parcel of our modern human nature. Saying "research shows" supposedly gives validity to stereotypic myopia, soundness to inane beliefs, authority for unilateral action, and protection from rampant stupidity.

Ideally, research tends to either dispel or put into perspective those differences—real or imagined—across such sensitive lines as gender, ethnicity, culture, and race that are often founded in stereotypes, prejudices, oversimplified generalizations, irrational belief systems, or old axioms, or derived from group pressures or disproportionate power bases. None of us wants to be judged based on stereotypes or prejudices. Any of us feel wronged when we or others whom we know are put down based on perceived—not actual—differences. Consequently, we all should welcome and applaud findings from research that put to rest any such unfounded and unequalizing voices.

A Bird Lost Forever

Two of your authors hail from North Carolina. At one time, there was a colorfully plumaged wild bird common to the Carolinas (actually, the Ohio Valley eastward) called the Carolina parakeet, a member of the parrot species. It was apparently similar to the multicolored parakeets seen in pet stores and once existed in large numbers. In the not-too-distant past (1904), people out walking the trails and forests along rivers were likely to see these colorful birds as they flitted here and there in the arboreal vastness.

The Carolina parakeet is now extinct (officially since 1939). Many times, as we are hiking through the North Carolina mountains, we try to imagine what it would be like to see such colorful birds above us. It is a sight we will never see, nor will our children or our children's children, down to the last generation.

While extinction is a natural process, we—that is, people—have contributed directly or indirectly to the demise of many species. Unfortunately, we seem to be leaving behind only those species that we, in particular, would most like to see become extinct, e.g., mosquitoes, no-see-ums, and roaches, all bountiful in the coastal regions of the southeastern United States. If given a magic wand that could be waved, thus extinguishing mosquitoes, etc., I am not sure we would wave it, though, despite the temptation, because somewhere in the back of our brains is the notion that we as a species will survive only as long as we can maintain viable biodiversity across the planet, even if that includes mosquitoes. We are afraid that much of that biodiversity may find itself limited to zoos in the not-so-distant future.

We are the top of the brain chain. Our religious and ethical roots teach us that we are to have dominion *and* stewardship over all creatures. We have become very good at the dominion part. We need to start stewardship *now*—more appropriately, last week or last year.

Will the human species die if the polar bear ceases to exist? Not immediately. However, the death of each species sets up a reverberation throughout this biome we have come to call Mother Earth. Eventually, we will feel the accumulation of those reverberations in our daily life and our daily survival. As much as we would like to ignore it, we are an interdependent species. Our "more primitive" ancestors, the American Indians, recognized this interdependence. Wouldn't it be great if we could recapture their respect for the sanctity and sacredness of Nature?

If that respect is to be recaptured and species saved from extinction for our children and grandchildren, it will be the scientists of today, through their research efforts, finding the paths to coexistence and re-educating us all on the beauty and necessity of what Nature has provided for us. It will be the scientists of today—and tomorrow and the day after—who grapple with what will be an ongoing squeeze between us and them, where "them" is any other species occupying a niche we think we are entitled to. If we are to recover our mandate to be good stewards of the Earth, it just may be research that leads the way.

Research and the Future

We as a species live at the top of the heap—or so we think. For better or worse, we have risen to the top despite our lack of long teeth, sharp claws, and speed. We have done so with our big brains. But not *just* big brains—rather, big brains driven by innate curiosity, a remarkable appreciation for self-protection, a growing love affair with comfort, and the seeds of the beginning of an awareness of our need to interact with each other and with our environment compatibly and with minimal damage to both.

As you read this chapter, perhaps even dreading this course, realize that you will walk the path of many researchers, some known and most unknown, who have put around you, sometimes even at the cost of their own lives, all the things you now have, from frivolous frills to your very immunity to deadly viruses. Even as you read this chapter, there are researchers, using standard research methods and statistics, working hard to corral bacteria that have outstripped many of our current antibiotics; design a safer, more fuel-efficient aircraft; predict major storms better; provide clean water. They are working out of curiosity, out of self-protection—and for you. To follow their footsteps along the research trail is, indeed, to tread hallowed ground.

To see how hallowed, look up the following individuals and contemplate for a minute how your life is different because of them:

- Ignaz Semmelweiz
- Alexander Fleming
- Jonas Salk
- Louis Leakey
- Wernher von Braun
- Edwin Hubble
- Daniel Burnham
- Frederick Law Olmsted
- Harry Harlow
- Mary Ainsworth
- Marie Curie
- Carl Sagan
- John Nash

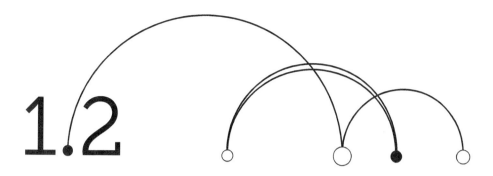

Defining *research methods*: How do they differ from statistics?

L et's clarify several terms that can be rather confusing: ***methodology***, ***method***, and ***design***. Are these important terms? Not necessarily in the everyday world of doing research. However, they are important as scaffolds on which we hang our understanding of research. It has been our experience that these terms are often used interchangeably, and that "apparent" interchangeability can be confusing.

What we want to do here is clarify what these terms mean, because they are very different, yet connected. First, we will differentiate among the terms and then we will reconceptualize them in a later section. This is a daunting task. The easiest way to go about this is to revert to the basic meaning of the terms *methodology*, *method*, and *design*. Here we go—hold on tight!

Methodology, Method, and Design

Research methodology

The term *methodology* is composed of two parts—*method* and *ology*. Like most *ology*-words, this implies the study of something. Methodology is the study of the methods or ways and processes of doing research. It involves looking at the broad principles that underpin research and the processes of doing research. These principles and processes deal with issues such as why we pursue knowledge, how we go about the pursuit of knowledge, what we do with the knowledge gained, and how knowledge feeds on itself by triggering additional questions based on answers found.

If you return to the very early part of this text, we suggested that mankind is driven to pursue knowledge because of curiosity and self-protection. This would be a fundamental part of methodology—the *why* and *how* of research.

Let us go at this from a slightly different direction and think in terms of basic and applied research. *Applied research* is designed to answer a question that has something to do with the real world. *Basic research* is designed to fill in gaps in our knowledge.

This distinction would fit the historical development of statistics, for example. In the late 1800s, a number of bright individuals were asking if there was a way to show mathematical patterns among variables. The inability to show patterns represented a gap in our knowledge. Anecdotal evidence and intuition suggested such patterns were evident in some relationships among variables, but were there ways to actually demonstrate this? People such as Galton, Dickson, and Pearson (Tankard, 1984) set out to see if this mathematical gap could be closed. They were seeking the basic mathematical methodology to represent patterns and relationships. When this gap was filled—when the methodology was developed—then it was applied looking at relationships among family members, kin, etc. The application represents the applied part.

As another example, Campbell and Fiske (1959) asked the question, "Is there a better way to demonstrate trait and instrument validation?" They were essentially suggesting that there might be a gap in our ability to do this mathematically. Through combined efforts, they developed an elegant methodology called the *multitrait-multimethod matrix analysis*. That methodology is now established as a method that researchers can use to answer specific questions about overlapping validation.

Note how we went rather seamlessly from methodology—the exploration of a gap in knowledge—to a method—the application of the methodology—with the Campbell and Fiske example. When scientists develop a new way of doing something—methodology—the new way quickly becomes a method to be used by other scientists.

Methodology also encompasses the reasoning behind the *how* of what we do in research. When you select a statistical method or a data collection method, there has to be a sound reason for those selections. Methodology provides the reasoning.

An example is in order. Let us suppose you are curious about whether there are plants on the Island of Nowhere that might prove useful in the treatment of many common diseases. How will you satisfy your curiosity? Where do you start? What is your first step?

A first step might be to determine which discipline and which knowledge base would be most useful to you in finding such plants. Certainly, you would not want to apply the knowledge base of astronomy to this pursuit. However, there is a body of knowledge out there that is most germane to what you want to do and you must find that body of knowledge. In other words, you must fill in that gap. That body of knowledge will also dictate, to some extent, general approaches that are consistent with that particular discipline that will help you gain the knowledge you are seeking. In short, and to use an analogy, you have to find out where you are first, before you can get to where you want to be.

In research, we talk about research methodology—the broad principles—and research methods—the application of the principles. *Research methodology* is how we go about conceptualizing research.

To reiterate, under research methodology, we might place things such as the why's of research, the how's of research as applied to different disciplines (biology, aerodynamics, chemistry ...), the goals of research (specific to disciplines), and the essential characteristics of research that make it research. Essential characteristics might include the advancement of knowledge, aligning your research with the appropriate discipline or disciplines, or the kinds of why's we can ask within that discipline. You might think of *methodology* as the blueprint for research, and *method* as the means of executing the blueprint.

We are not sure that we have clearly elucidated the difference. It is a fuzzy distinction with which to wrestle. Perhaps an example not related to mathematics would help.

Let us try skiing as that example. We know that skiing involves gravity (as you know very well if you have skied), weather conditions, powdery white stuff, etc. These are the overall conditions necessary for skiing. Somebody at some point in time had to wonder if it might be possible to slide down that white slippery stuff on boards. Could it be done? That was the gap in knowledge. Perhaps a number of times spent sliding downhill, whether by design or accident, on one's derriere gave impetus to the investigation of possibly another and better mechanism for sliding than one's backside. The result? Skis!

Now we can move into the applied process. Your authors learned to ski by the French-Swiss method. Note the word *method*. One of our wives was taught to ski by her father. He put her on skis, took her to the top of a mountain, pointed her downhill, and said, "Go." That was the "Better learn quick" method. In either case, a method was used to teach someone to do something, but only after the gap in knowledge had been filled in.

To reiterate, method is a way of doing something once methodology has answered the questions of why and how.

One more example, if you will bear with us. In making steel, there is a methodology that involves chemistry, physics, geology, lots of heat, etc., which tells us the general conditions for making steel, ingredients needed, the physics and chemistry of the process, etc. We then have to adapt those general conditions to the actual doing—the method—the actual making of the steel. When the adaptation has been completed, we have the methods for making steel—the oxygen method or the arc furnace method. (There may be more.) The point is that these two methods represent the application of the theory (methodology) of making steel.

Methods are simply the adaptation of methodology to the actual doing—in this case, the act of making the steel. It is the same with almost anything. There is a methodology that outlines broad parameters and methods that put those parameters into action.

Research method

Method, then, is a way of doing something. In the steel industry, there are methods of making steel. In skiing, there are methods of teaching skiing. In statistics, there are methods of finding answers to research questions. Looking for a relationship between variables? There is a method. Trying to predict tomorrow's weather? There is a method. Looking for similarities and differences between the genders? There is a method.

As a statistical example, let us look at the issue of possible differences between the genders. More specifically, you want to know who spends the most on lunches—men or women. You will want to use an inferential method to answer this question. Inferential methods help us isolate dependable findings about behavior, find answers that can be extrapolated to the population, or establish causality. Had the research question been different, you might have used a descriptive method, or maybe a predictive method.

Research method—the "doing" part of research—has several important steps. The first thing you need to do is decide what you are interested in investigating. There are a lot of topics out there. Which one interests you? Given that some research can extend over a prolonged period, it really does help to be looking into something that you find interesting. While the most important point, from a student's perspective, is probably just to get the research project done and get a grade, it really helps to accomplish that if there is an intrinsic interest to start with.

Once you have identified an area of interest, you have to start thinking about the research question that you will pursue. A good way to do this is to find an introductory text that covers your area of interest and read through that part of the text. Introductory texts are good because they survey such a broad area of research in specific areas. It will give you an overview of what has been done and what remains to be done. You might think of this as a preliminary literature review. By the time you finish this, you will know whether—or not—your research question is a new one, if it replicates existing research (which is fine), or if it is even doable in the timeframe that you have.

There is an old saying known as the KISS principle: Keep It Simple, Stupid. Not very complimentary, but very *apropos* for research. Too many projects have been bogged down in questions that were overly weighty or required years of effort.

The research question will identify the target of the research (people, mice, monkeys, quarks, chemicals, etc.). The target defines how you will collect the data (and there are a number of methods for data collection). The data collection determines how you will analyze your data and, finally, how you will present your findings. This is a somewhat rigid sequence that researchers follow in conducting research.

Research across most disciplines follows this pattern. We know of no discipline that goes out and collects data before deciding on the research question, and no discipline identifies the research group before deciding on the research question or method of collecting data. That would be like starting at the end and working toward the beginning. Thus, at a general level, *research methods* refers to the overall pattern of identifying an area of interest, asking a question, collecting data, analyzing the data, and presenting the results. This process will have many features to it that are discipline-specific.

The concept of research methods, we hope, is clear. It is how you go about implementing research—in our case, producing a research study—rather than producing steel or teaching intrepid skiers.

Unfortunately, *research method* often overlaps or is used interchangeably with *research design*. Let us elaborate on our previous examples. The French-Swiss ski method has a design to it: short skis to start with, no ski poles so you do not stab yourself or others, etc. … The method for producing steel has a design to it: using heat to melt ore that is separated from slag. Design is the fundamental piece that gets you to the nitty-gritty of the doing.

Here is a brief scaffolding of what we have just covered that may help:

- Research Methodology: The study of research methods
  - Goals and objectives of research
  - The furthering of knowledge
  - Discipline-specific considerations
  - Development of new methods
- Research Methods
  - The order of research investigation
  - Area of interest and development of research question
  - Subject identification and data collection method
  - Selection of analysis technique(s)
- Research Design (discipline-specific)
  - Statistical analysis technique
  - Analysis software interpretation and presentation

Research designs

Think of it this way: If *methodology* involves the overarching principles/essentials of the research process, and if *method* brings methodology down one level to pragmatic considerations of how you are going to do something, then *design* represents the nitty-gritty of actually finding folks for the study, measuring and collecting data following some prescribed format, recording the data to be compatible with a computer routine, selecting a statistical analysis procedure, etc.

Within that nitty-gritty, a basic set of six research designs is frequently found in the literature of most fields. These designs are by no means limited and can be applied to both the social sciences and the hard sciences such as physics or chemistry or biology. The six "common" designs are:

1. **Experimental designs**—how subjects are selected and assigned to treatment or laboratory conditions; the "how we handle subjects" part of doing experimental research
2. **Quasi-experimental designs**—how subjects are selected and assigned when a quasi-experimental variable is involved
3. **Correlational designs**—the mathematical relationship between two variables
4. **Survey designs**—using surveys for the collection of data
5. **Single-subject designs**—involve a single subject
6. **Qualitative designs**—those that do not involve any math and are simply qualitative in nature

Designs other than statistical may be discipline-specific.

Note that these statistical research designs are the third layer down in the brief scaffolding shown previously. We can look to Kerlinger (1986), who suggested that a research design is an overall plan and structure aimed at answering some research question, while minimizing extraneous variables, reducing error, and maximizing variance (more about these later). We think this is an accurate definition of research designs. Each of these designs has ways of controlling extraneous variables, minimizing error, and maximizing variance, which are part of the nitty-gritty of research.

Here is more about the basic six.

Experimental designs

An experimental design is how a researcher selects and assigns subjects randomly to treatment conditions. Here is an example to help illustrate this.

A number of years ago, one of the authors and a colleague developed a simple research study that was designed to demonstrate the effects of sleep deprivation on motor control and speed. A study on sleep deprivation could be used to examine subjects in many fields, such as pilots, long-haul truck drivers, surgeons, etc. We asked for volunteers from among our graduate students and assigned them randomly to one of three treatment conditions (the independent variable) as follows:

- Treatment condition 1, the control group—subjects were to sleep for eight hours each night for three consecutive nights.
- Treatment condition 2—subjects were to sleep for six hours per night for three consecutive nights.
- Treatment condition 3—subjects were to sleep for four hours per night for three consecutive nights.

In this experimental design, the amount of sleep the subjects were to receive was the **independent variable**; that is, it was under the control of the investigators. Having established the independent variable, we then needed a measure that might be sensitive to sleep deprivation. We elected to use the Finger Oscillation Test (FOT) from the Halstead-Reitan Battery. This test requires subjects to tap a telegraph-like key with the index finger of the dominant hand for one minute as fast as they can. A counter tabulates the number of taps. It is a fine-motor test, and we felt it would be sensitive to disruption in speed of taps as sleep deprivation increased, thus demonstrating the effects of sleep deprivation on fine-motor ability. It also would be infinitely safer than sending potentially sleep-deprived students out to drive as our measure of the effects of the independent variable.

The FOT, then, was our **dependent variable**—it relies or depends on the performance of the subjects. The design, presented graphically, would look like this:

8 Hours	6 Hours	4 Hours
Finger oscillation mean	Finger oscillation mean	Finger oscillation mean

We obtained a baseline on each subject by averaging the number of beats per minute over three trials, with a two-minute rest between trials following three nights of eight hours of sleep for all subjects. We then started the sleep deprivation phase of the study for the next three nights. At the end of the three nights of sleep deprivation, we readministered the FOT.

This was a typical experimental design in its simplest form. It can be recognized by its levels of treatment conditions and by the random assignment of subjects to the treatment conditions. Because we did not have a large number of graduate students, we were not able to use random selection in addition to random assignment. Use of random selection with random assignment would have been ideal.

Consequently, one might say that an experimental design is recognized by random assignment of subjects across predetermined control and/or treatment conditions utilizing a dependent measure to assess the manipulation. These are its distinguishing features.

Quasi-experimental designs

For quasi-experimental designs, think experimental design with a quasi-experimental variable under study. A quasi-experimental variable is one that must be pre-existing in subjects because, by the nature of the variable, a researcher cannot ask subjects to assume the behaviors and/or risks that would go with the variable since the variable itself might be potentially harmful.

Examples of this would be depression or smoking. You cannot ask nondepressed subjects to become depressed or nonsmokers to start smoking just to participate in your study of depression or smoking behaviors. However, we can study subjects who are already depressed or who already smoke.

As an example, let us suppose that we want to study the effects of various treatments on depression. We might design three treatment conditions—again, as in an experimental design. (Although there is nothing magical about having three treatment conditions; there could be fewer or more.)

Treatment condition 1 might be psychotherapy only. Treatment condition 2 might be psychotherapy along with appropriate pharmacological (drug) treatment. Treatment condition 3 might be psychotherapy along with a consistent and monitored moderate-exercise program. Note that there is no control group, which would be a no-treatment group, since it would potentially be unethical to place depressed individuals in a no-treatment condition.

Next, we must find a population of depressed people. We then randomly select (if we can) and randomly assign already depressed individuals to one of the three treatment conditions (with their consent, of course), providing they are not already in treatment. This is an experimental study but uses a quasi-experimental variable. We can see that a major distinguishing characteristic of this kind of study is that the subjects bring pre-existing conditions with them (the quasi-experimental variable).

Correlational designs

Correlation is a statistical technique credited to Sir Francis Galton (Charles Darwin's half-cousin) and presented in his 1888 paper to the Royal Society of London entitled "Co-relations and their measurement chiefly from anthropometric data" (Tankard, 1984). The technique has been refined over time by others and generally bears their names (e.g., Pearson, Spearman), rather than Galton's.

The technique examines the potential for relationships between variables that might logically seem to be related, such as intelligence and grades, fluctuations in the length of women's hemlines and stock market activity (yes, there is a relationship), etc. It simply identifies a mathematical relationship. It does not, in any way, establish causality.

We might say that correlational designs are recognized by the statistical examination of the relationship between two variables that seem as if they should be connected logically.

Survey designs

Survey designs use surveys or questionnaires to gather data. A survey or questionnaire is generally a paper-and-pencil method of asking questions and having subjects respond to the questions, either writing open-ended responses or choosing among offered responses (multiple choice), or something in between. There is a proliferation of surveys on the World Wide Web; one has only to visit sites such as www.surveymonkey.com to see a number of different survey-type instruments on a broad range of topics.

Single-subject designs

A single-subject design uses only one subject. As an example, one of the authors worked for a number of years with a colleague at an inpatient psychiatric hospital for children, where a young woman on one of the wards had a particularly recalcitrant case of trichotillomania (pulling out one's hair). A rather innovative behavioral intervention was successful, and the colleague wrote up the intervention for publication. The published article would represent a single-subject design.

Qualitative designs

Qualitative designs are the opposite of quantitative designs, which we have been talking about so far. They do not include mathematical data and are, instead, based upon nonmathematical observations or words. This type of research is seen often in anthropological research, such as Jane Goodall's groundbreaking work with wild chimpanzees. Qualitative designs are recognized, therefore, by the lack of numerical data.

Note that each design can be defined by a very narrow parameter: subject assignment (experimental), type of variable under study (quasi-experimental), statistical technique (correlational), data collection technique (survey), using only one subject (single-subject), and absence of numerical data (qualitative).

As you can see, these designs are quite specific about how the researcher handles subjects (or a subject), what kinds of data are collected (quantitative or qualitative), potential data collection techniques, etc.

Statistics

Before getting to our reconceptualization of research methods and design in the next section, we need to put statistics in its place, so to speak. Where do statistics fit into this scheme?

Let us offer another analogy at this point. You are a contractor and you build houses. You have a house being built right now, in fact. You have some masons onsite, laying a stone wall around the patio in the backyard. They are splitting and shaping the stones using a mason's hammer. Up on the roof, more workers are laying down an asphalt shingle roof. They are using roofing hammers. Around the side of the house, carpenters are doing interior work on joints, studs, and such, and they are using carpenter's hammers.

Are you getting the picture? You use different hammers for different tasks.

Similarly, you use different statistics for different research questions and/or research designs. Statistics, like hammers, are just a tool. They do not define anything. They do not dictate any part of the methods or design. They sit by, idly waiting until you have data that address your research question. Then they jump in and pound a nail, rip up a piece of roofing, or split a rock, depending on what you need done. Simple, isn't it?

Where to Next?

The next section presents what your authors feel is a generic hierarchy of the methods of research that fits almost any discipline conducting legitimate research. You will see that the traditional designs described here will fit into this hierarchy and that the basic components we discussed under research methods are represented as well. We think you will like this approach.

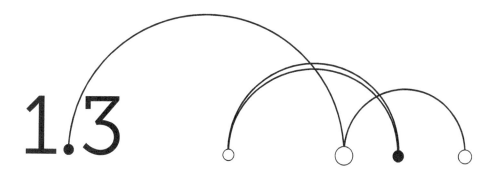

1.3

A conceptual framework for research methods

There are entire books devoted to describing research designs and research methods. If you read a book, particularly one on research designs, you will find several designs, as mentioned in the last section, that constitute the "basic" research designs. These would include experimental, quasi-experimental, correlational, survey, qualitative, and single-subject designs.

Your authors do not think that these designs rise to the level of "research designs." While these may be designs in a loose sense, they are identified primarily by more mundane considerations, as already mentioned—that is, experimental designs are recognized (defined) primarily by their subject-assignment procedures (random assignment to multiple treatment conditions); quasi-experimental designs primarily by their use of quasi-experimental variables (variables that subjects cannot be randomly assigned to due to health and/or ethical considerations—you cannot assign a nonsmoker to a smoking group, thus making smoking a quasi-experimental variable); etc.

From your authors' perspective, how you arrange data or the statistical methodology used to analyze data should not be the primary determining characteristic(s) of your research.

To put all of this in place, we offer a research methods hierarchy that covers all disciplines. There are five levels to our generic hierarchy. If you are familiar with Abraham Maslow's hierarchy of needs, that may help you form a conceptual base for the type of hierarchy we are going to create. The hierarchy we are advocating looks like Figure 1.1.

FIGURE 1.1 Generic Hierarchy

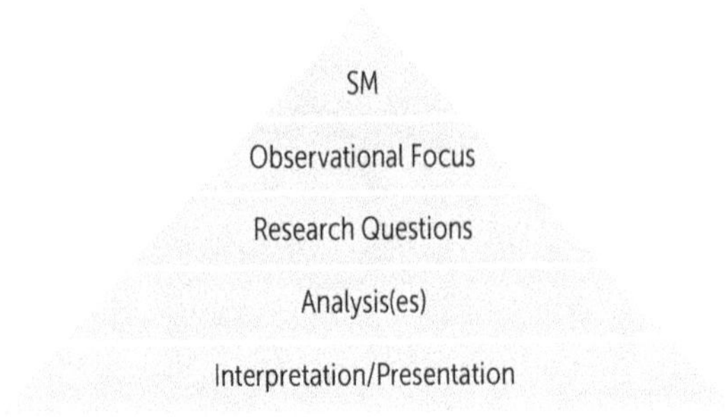

SM

Observational Focus

Research Questions

Analysis(es)

Interpretation/Presentation

Level I: The Scientific Method

At a very broad level, we are interested in how we come to know what it is that we think we know—our method for acquiring information, one such method being research. Peirce (Weiner, 1966) suggested three methods of knowing. One is the tenacity method. In this case, we have heard something repeatedly or have had a particular or unique experience. At this level, our belief stands no matter what. We cling to the belief with tenacity—regardless of whether it is supported by reality or data.

Another method of knowing is derived from authority. This is certainly used by Madison Avenue in commercials. How many times have you seen an individual dressed in a white lab coat, for all appearances seeming to be an MD, touting the virtues of some medicine? Advertisers are not dumb—they are presenting you with an authority figure who is telling you what drugs you should be asking your personal physician about for your particular ailments.

Another method of knowing identified by Peirce is the a priori method. This one is based on logic and reasoning. While it is more sophisticated than the other two, it is still subject to error (excluding, of course, Mr. Spock), should your data be faulty or your logic illogical.

There is also the empirical method of knowing (Turner, 1967). This method simply makes use of objective and verifiable observations or measurements to support what we know.

The last two ways of knowing give rise to the **Scientific Method (SM)** (Turner, 1967). The Scientific Method sets forth the tenets of research and thus provides us with the top and most-general part of our hierarchy.

The Scientific Method states the basic guidelines that support all "research"—all legitimate research, that is. It starts with, primarily, three tenets, and then goes on to describe the relevant characteristics of data and of researchers. If you look back over the last section, you will find some of the elements of research that will appear in the discussion of the Scientific Method. Let us start with the three basic tenets.

The first tenet suggests that Nature is lawful. To assume otherwise is to suggest that Nature is random, and random events cannot be studied systematically because they *are* random. The second tenet suggests that the laws that underpin Nature can be identified and understood through systematic investigation (research). Again, the counterpoint is that, if the laws cannot be understood, even if lawful, then research would be fruitless. Finally, the third tenet views behavior as deterministic; that is, influenced in a cause-effect manner by internal and external events that can be identified. This would include the behavior of humans, as well as planets, molecules, etc. In particular, humans like to think that we are agents exercising free will. Let us debunk that myth by asking one simple question: Are you reading this text out of free will?

You may have problems accepting these three tenets. Philosophers have argued back and forth for centuries about the lawfulness, "understandableness," and deterministic qualities of Nature, so you are not alone if you ponder the veracity of these three tenets. It is not our purpose to argue the finer points of these assumptions—only to suggest that, for research to be conducted and move ahead, these assumptions must be embraced at some level.

The Scientific Method also provides guidelines for data collected for study. That is, data should be empirical (measurable, rather than nonmeasurable), objective (rather than subjective or metaphysical), systematically gathered (rather than haphazardly so), controlled (to eliminate competing or extraneous explanations or factors), and verifiable by others (the researcher has described the objective and systematically gathered data, and the controlled conditions under which they were gathered, well enough for others to duplicate and verify—or not—the findings).

In particular, verification is an important element in validating previous research findings, catching research "errors," and identifying false research. Imagine how the lead article in that newspaper you see at the check-out counter at the grocery store, "Martian Aliens Spotted at Saks Fifth Avenue in Los Angeles," might hold up against the application of the tenets of the Scientific Method or the characteristics of data as described above. Are Martians objective? Can data be gathered about them systematically while shopping at Saks? Would a study on Martians be verifiable? Certainly raises some questions, does it not?

The Scientific Method also details relevant characteristics of individuals conducting research. Researchers are assumed to be uncertain (aware that they do not already know everything), open-minded, skeptical (always questioning research findings, looking for flaws or methodological weaknesses), cautious (in declaring any research as having found the truth with a capital "T"), and ethical. This last point is a bit of a stickler and has led to some embarrassing moments in the history of research. A slight digression, to briefly outline a few of those embarrassing moments, might be in order.

Piltdown Man

In 1907, some bone fragments were found in Heidelberg, Germany, that suggested that the earliest known human ancestor was German, thus making Germany the birthplace of the human species. Other European countries vying for "first" were chagrined, and thus were laid the seeds for fraud.

Charles Dawson, a lawyer and amateur paleontologist, was given some skull fragments found by a gardener in Piltdown Common in Sussex, England, in 1908. When more bone fragments were discovered, Dawson took them to Dr. Arthur Smith Woodward of the British Museum of Natural History. Both men, along with Teilhard de Chardin, a friend of Dawson, returned to dig in the area, finding even more bone fragments. Indications were that the age of the bone fragments topped the fossils found at Heidelberg that had established Germany as the birthplace of the earliest precursor to modern man. Woodward assembled the fragments back at the museum; they indicated a human-like skull and ape-like jaw bone. The find became known as the "Dawsoni Dawn Man" or just "Dawn Man," as well as "Piltdown Man" in reference to where the remains were found, and the English laid claim to the earliest human ancestor. In 1953, sophisticated fluorine analysis exposed the Dawn Man as a fraud.

Who perpetrated the fraud is not clear. It could have been Dawson, Woodward, or de Chardin, or even somebody else. Nobody really knows, although even Sir Arthur Conan Doyle's name often comes up in this regard—playing a prank on either or both men, or mankind in general. Regardless, foundational guidelines of the Scientific Method, particularly objectivity, verification, and ethical behavior, were violated, and the impediments imposed on continued research and exploration by this fraud over the period from its discovery to its proven falsification are incalculable.

de Chardin, left; Dawson, seated, and Woodward, right, at Piltdown Commons. Would the Scientific Method find that was the Aflac duck, lower right, with the investigative team?

Sir Cyril Burt

Sir Cyril Burt, who died in 1971, was a professor of psychology at University College, London, who had been knighted for his work in genetics and received the prestigious Thorndike Award from the American Psychological Association.

His research centered primarily on the clarification of "nature" and "nurture" contributions to development and was based on a large research cohort of monozygotic twins separated at birth, thus varying nurture while holding nature constant. His research on intelligence suggested that upward of 75% of intelligence was inherited, lending scientific "fact" to the eugenics movement and theories espousing genetically based intellectual differences among races.

Leon Kamin, then at Princeton University, raised some questions concerning the veracity of Burt's data, but this was dismissed by the British psychological establishment as mere jealousy among American psychologists of Burt's professional stature.

Leslie Hearnshaw, professor of psychology at the University of Liverpool and the eulogist at Burt's funeral, was commissioned to write a biography of Burt's life. Imagine the impact when it was found that Burt perhaps did not have empirical, objective, or verifiable data. There were indications he had made up some, if not all, of his data. The research assistants who supposedly collected the data could never be found. Imagine the impact of his "research" findings on the nature-nurture debate relative to intelligence during those years he was falsifying data!

Level II: Observational Focus

What each researcher looks at in terms of the observational focus is a function of the researcher's discipline. Physicists, for example, may select subatomic particles as an observational focus for research. Historians may select historical documents, astronomers may select galaxies, and so on. Psychologists most often select people, although rats, monkeys, and other animals may form a suitable substitute. (It is often said in psychology that everything we know is based on rats and college freshmen, and the data from rats may be the most reliable.) The point is that the material selected for observation is generally discipline-specific.

A few years ago, one of your authors was sitting in the dentist's office awaiting a dental exam and cleaning (an event illustrative of the Freudian notion of oral aggression) when an article in *American Heritage* dealing with the death of Abraham Lincoln caught his eye ("How did Lincoln die?" by R. A. R. Fraser, MD, *American Heritage*, March, 1995, pp. 63). The basic hypothesis of the article was that, if the doctors in attendance had just left Lincoln's head wound alone, Lincoln most likely would have survived. Interesting article! Here is what Dr. Fraser did.

The derringer used by John Wilkes Booth was examined and projectile velocity calculated based on projectile weight, barrel length, powder composition, and amount of power used. Diagrams of the bullet path drawn by the doctors were examined to determine which parts of the brain had been injured. Modern records for similar head gunshot wounds were examined for survival rates. Bottom line, the survival rate is quite high with these types of wounds, e.g., low-velocity bullet entering the cortical area only. The constant probing of Lincoln's brain by the doctors with nonsterilized fingers introduced infection, caused continued bleeding, and perhaps caused additional damage to brain tissue. Did these doctors know better? An examination of medical school textbooks for the 20 years before the assassination taught that the best practice with head wounds was to leave the wound alone after clearing the surface area of bone fragments, etc.

Dr. Fraser's conclusion? Had Lincoln been left alone and medical protocol been followed, he may well have lived, although he might have had some degree of impairment, perhaps much like Congresswoman Gabby Giffords (p. 63).

Here was an interesting article from the field of history that followed the guidelines of our hierarchy as applied to materials that we do not normally think of when we think of research. This article involved good research, carefully conducted and clearly explained, and can be examined by others and repeated for verification. The conclusions are drawn from the findings and presented not as truth with a capital "T," but as a potential logical conclusion based on the evidence.

Level III: The Research Question

For any particular discipline or research study, only three viable research questions can be asked:

- How does it look (descriptive)?
- What caused or influenced it (inferential)?
- Will it happen again (predictive)?

Describing, inferring causality, or predicting is about the limits of what research can demonstrate, regardless of the discipline. Usually, each individual research study is geared toward only one of these questions, although that is not a hard-and-fast rule—some studies may address more than one question at a time.

Predictive issues are generally arrived at after a number of studies have identified, descriptively, relationships among the phenomenon under investigation.

The tendency is to ask the descriptive questions during the early part of an investigation of some new area, move to the inferential questions as more information is learned about the area, and possibly move to predictive questions when influencing factors have been isolated, providing, of course, that prediction serves a useful purpose.

Volcanology is a good example of this progression. Initially, volcanologists were primarily interested in where active volcanoes were located—mapping the terrain and **describing** locations. Having found the volcanoes, there was further investigation into descriptive factors: how often do they erupt, how long does an eruption last, what different kinds of eruptions are there (from blow-outs to oozes), etc.

The next step was to use inferential studies to start looking for **causal** factors, such as stress on the tectonic plates. Now the attempt is to take those causal factors and see if they can be used to **predict** such eruptions. Obviously, prediction is the most difficult part of the process, but clearly the end goal when dealing with a potentially destructive force like volcanic eruptions. That prediction is the most difficult use to make of any database from any research field can be seen simply by watching the daily weather report.

Some suggest a fourth research question that deals with control. Certainly, behavioral principles derived from research have been used in clinical settings to control and modify individual behaviors. The issue of control as a research question really involves combining causality and prediction, and represents a logical extrapolation from such research. In other words, causal factors are identified from the research and applied in a systematic way to an individual or an event, with careful observation and measurement of the behavior or the event to the systematic application of the identified causal factors.

A change in the behavior or event under highly controlled circumstances in the direction predicted by the systematic application of the causal factors to the behavior or event is then seen to confirm the influence of those causal factors on the behavior or event. Then comes the extrapolation which, simply put, goes like this: If A causes B, then A may control B to some extent.

A cautionary note: While the research questions may seem simple, answering them may not be. It may take years of data collection and analysis just to begin to describe a phenomenon clearly enough to move into more causal kinds of research. This long trek toward beginning to fully understand an area is illustrated by the extraordinary amount of time it might take to get a new drug to market, the ongoing study of intelligence (now well into its second 100 years), the ongoing study and prediction of hurricanes and the much-younger field of tornado prediction, etc. Research progresses by small steps. Rare is the research study that sets a field on edge and redefines the direction of research.

Level IV: Data Collection and Analysis(es)

Once we know what we are studying (observational focus) and have zeroed in on a relevant research question, we collect and analyze our data. In other words, if the research question is the "what" of our research—what we will be looking for—then the analysis is the "how" of our research question: How will we look for what it is that we are looking for; that is, how will we collect and examine the data to address the research question posed?

Both data collection and data analysis are critical to the research question. Most basic is the issue of data collection. Data collected incorrectly will negate any research question even if the analysis can still be made. Conducting an analysis with flawed data is fraught with problems, apart from the issue of the flaws in the data themselves. We will deal more with the issues involved in actual data collection in later sections. Suffice to say, at this point, that this is a critical process in the overall success of your research.

How data are analyzed is also important and will be driven by both the research question and the method of data collection (this is important—you need to think about what it means), and may well be particular to the specific discipline asking the questions and collecting the data. In disciplines such as psychology, sociology, and political science, statistics may be the primary "how" of data analysis. For other disciplines, the "how" may involve quantum mechanics (physics), authentication of handwriting through analysis (the discipline of graphology often used in history and literature), measurement of wide bands of light refraction from stellar bodies (astronomy), etc.

This is not to say that statistics cannot be used in any of these disciplines. It is to say that it does not *have* to be used to confirm that research is taking place.

Going back to our earlier comment about designs, we now see where statistics comes into the picture. Statistics is simply a "how"—a tool, one part of the overall hierarchy. Contrary to many traditional methods of describing designs, your authors feel that the statistical procedures are only one small (but important) part of the design and not *the* defining characteristics of the research design itself. Really, statistics is like the hammers mentioned earlier—simply various techniques for "hammering" data.

This is not to undermine the importance of data collection and analysis, but only to say that, in our conception of research designs, this constitutes only one level of the overall hierarchy.

Level V: Reporting and Interpretation

Having completed the research from the assumed adherence to the Scientific Method through selecting the observational focus, to clarification of the research question, through data collection and analysis, the researcher must now interpret and report those results. This is the final component of our hierarchy.

In psychology and like fields, this involves reporting the results of any statistical analysis(es) of the data and then placing those results in the stream of research that has preceded the study with an explanation of what the findings suggest and how those findings clarify, add to, or redirect previous research. In good studies, your research will advance the knowledge in the area by a step—perhaps a small step, but then, small steps are still valuable.

"Groan, groan," say students doing research

Many master's-level students complain about being required to do a thesis (doctoral students sometimes complain about dissertations, but more quietly). Nowadays, many American universities also require a final research capstone at the undergraduate level. Students perceive the requirement of a research capstone or thesis as punitive action by faculty. Some students have suggested that their studies will be so nondescript as to fade into meaninglessness, so why do it?

One cannot argue with that logic. The subject of a thesis or capstone may be a study that is so small, so inconsequential, that it does border on meaningless. The real meaning of a thesis, however, may not be the magnitude of the results. The real meaning may simply be doing it. What is it like to develop a research question? How do you decide where your subjects have to come from to adequately address your research question? How do you collect data that will be relevant? What magical statistical techniques will you use to bring your data to life? What have you found, and is it really meaningless (or meaningful)?

On completing the thesis or dissertation, there is at least the appreciation of what it took to follow through from beginning to end, regardless of how others may feel about the importance of the study. To the extent that a student learned how to conduct research, from beginning to end, that study can never be meaningless, regardless of the findings. Let us face it: Not all research can be earth-shaking—but all research can be meaningful to the person who had the idea and the perseverance to follow through to completion of the study.

Completing a study or capstone is also a testament to your perseverance. You started something and then finished it—no small accomplishment! One of your authors, in reading through a bulletin, found an advertisement by an individual with the letters "PhD ABD" after her name. There is no such thing. You either did it or you did not—ABD (All But Dissertation) simply means you quit before you finished.

This probably means we could add another relevant characteristic to those of researchers mentioned at the beginning of this chapter—the individual discipline to follow through to the end of a study. There is always the possibility that your study, no matter how nondescript you or others may think it is, will trigger ideas that lead to further research, which, in time, may lead to very meaningful findings. That, then, would mean that your thesis or capstone study was very meaningful.

In yet another way, doing a research study is like eating carrots. Remember how your mother used to say, "How do you know you don't like carrots if you never try them?" It's the same for research: How do you know you do not like it if you have never done it?

Where Does All This Leave Us?

Ideally, the material in this chapter has left you, so far at least, not confused, but aware that a research study is a complex configuration and integration of the components covered so far, and aware that statistics do not totally define the design. We hope you now have some structure in your head relative to the overall process and steps involved and are raring to go on your own research journey.

In Sum

If we could summarize at this point, we have learned that our hierarchy is composed of the Scientific Method as an always-assumed element. We have learned that the research question drives the research, that data collection is critically important for reasons yet to be disclosed, that statistics are a "how" and much like a hammer, and that there is a great deal of similarity between research and carrots.

We hope that you have learned that research can be as varied as anything else, such as houses. One of your authors drives by a small housing development on periodic trips to the western North Carolina mountains. Structurally, each house is exactly the same. The only difference is the color of paint. That difference in paint makes each house unique, despite the structural similarities.

We could do a study looking at the correlation between mother and infant intelligence. We could do the exact same study in all respects, switching fathers for mothers, and we would have a new study. It only takes a tweak here and a tweak there to change a study, thus creating an infinite possibility of different investigations. How many variations of research designs are there? As many variations as there are in humanity's ability to think up tweaks.

Where to Next?

For our next step, we will take a closer look at the "research question" part of the hierarchy. How do we form questions? How are they stated? How do questions direct the research? How do questions influence the statistics we select?

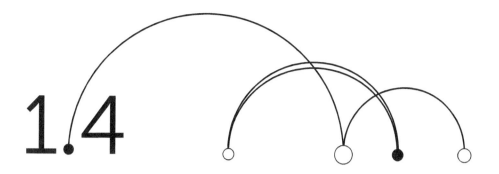

1.4

Research methods flowchart: Your map to understanding

As you will see throughout this book, we are very fond of flowcharts. They can be effective visual aids in helping you see the overall process of research. We present here a very short flowchart that shows the initial considerations for conducting research.

FIGURE 1.2 Qualitative, Quantitative, or Mixed Methods

Identify area of interest → Research question

Numerical/causal-inferential predictions → Quantitative (See Chapter 5)

Why, how, what Experiences, perspective, beliefs, feelings (See Section 6.2) → Qualitative (See Chapter 6)

Mixed methods
Hybrid method of qualitative and quantitative methods

Notice that we start the flowchart with identifying an area of interest. We mentioned this earlier and suggested it was important, since it provides some intrinsic motivation to doggedly pursue the goal of completing the research.

Out of that area of research interest will come your research question. Your research question will be quantitative or qualitative, or perhaps a mixed design combining quantitative or qualitative. You will make this determination by asking what kind of data you have: nominal, ordinal, interval, or ratio (more about these later). If you have interval or ratio data, you will most likely head in the direction of quantitative methods; if not, you will go in the direction of qualitative methods. The distinguishing feature between the two is whether you will have numbers (quantitative) or not have numbers (qualitative) to deal with.

Notice that on the quantitative side, you head off to look at causal or inferential (or predictive) issues using numbers. On the qualitative side, you look for "how," "why," and "what" kinds of answers, or exploration of more subjective states such as experiences, perspectives, feelings, beliefs, etc.

Later we will look at quantitative methods as one track that a researcher can follow. At that point, we will present you with this same flowchart, expanded to cover quantitative considerations. We will do the same in the qualitative chapter.

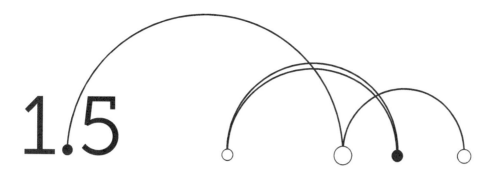

1.5

Research questions:
The driving force

Research is most clearly driven by the **research question**—what we want to know. The research question is a direct function of the content area of the particular discipline. Looking at Figure 1.1, note that the observational focus is the second tier down and the research question is the third tier. These two components—which are really inseparable—constitute the directing force behind research.

Again, the observational focus is discipline-specific. In psychology, it is often people, rats, or even primates. In chemistry, it might be chemical interactions; in physics, the path of particles, or the dimensional vibrations in the gravitational field measured by interferometers. History majors might study old documents; English majors might study poetry; art historians, art. The observational focus is truly discipline-dependent.

The research question derives from the observational focus. The actual questions asked are more similar to each other across disciplines than are the various disciplines themselves, since there is a limit to the types of questions that can be asked.

Types of Research Questions—From the Silly to the Mundane

The research questions boil down to "what," "how," "why," "when," and "where" questions. We might ask what something looks like (What does marriage look like after 20 years? What does the infrared spectrum of an imploding star look like?). These kinds of questions are pretty much descriptive.

Descriptive questions are very important, particularly in a new area of research where you are trying to find out what the lay of the land looks like. We might ask "how" and "why" questions when we are trying to pin down relationships in our data that indicate causal influences across variables or events. We might ask:

- How does a selective serotonin reuptake inhibitor (SSRI) influence depression?
- Why does an SSRI work?
- How do we negate flashback events from PTSD?

"When" and "where" questions are aimed at trying to pinpoint locations or times in many instances. We might ask:

- When did sexual mores change in American culture?
- When did Pangaea break apart?
- When will the next Pacific Rim earthquake take place, and where?

These research questions generally come in two varieties, as can be seen above: descriptive and inferential, matching to some extent the mathematical and statistical methods of inquiry described earlier. Some of the proposed research questions listed above are clearly descriptive, some more inferential. Some research questions might be labeled as more predictive, such as the ones dealing with Pacific Rim earthquakes and the timing of such events.

In many fields of research, the final goal really is prediction, particularly in regard to potential natural disasters such as hurricanes or tornados, or in manmade disasters such as a Columbine-type situation during the course of a school year.

These predictive questions are certainly relevant at a pragmatic level, but stretch the bounds of research findings and actually represent more of an application of educated conjecture based on research data than fine-tuned questions that can be answered directly by the research data. The best we can do is establish probabilities of occurrences in a broad, general sense, given sufficient data and information about the relationships of key variables that impinge on the predictive question being asked.

Even so, the science of prediction has advanced considerably, so we now have days, if not weeks, of advance warnings, for example, concerning the strength and path of potential hurricanes. We only have to read about the Galveston, Texas, hurricane of September 8, 1900, to realize how far we have come in terms of prediction. That single hurricane holds the record for the greatest loss of life in the United States from any one event.

The probabilities associated with some future event—as difficult as they are to determine—are likely to be more accurate in the physical sciences than in the sciences dealing with human behavior. When the forecast comes out for the East Coast hurricane season, your authors take note and take heed—there is some solid science behind these predictions. When someone says they can predict what another person is going to do a month from now, year from now, etc., however, we take it with less seriousness.

There is an old saying in clinical psychology that the best predictor of future behavior is past behavior, and a psychologist who lets him- or herself be led into making predictions about a specific individual's future behavior in a court hearing is opening the door for a very rough handling from the opposing side. Psychologists are little better, if at all, than psychics at predicting future behavior—and we all know how effective psychics are in predicting the future for movie stars, political figures, or world events.

Characteristics of Good Research Questions

Research questions should have certain characteristics to be *good* research questions. They should be **parsimonious**. That simply means that they are very carefully stated—even over-carefully stated—and kept simple. One definition of parsimony is "stinginess," so what we are talking about here is a stinginess of words.

Research questions also should be as **precise** as possible. Precision focuses the research question—cuts away competing chaff—and increases the viability of the investigation of the question by stating, without ambiguity, what one wants to know. The research question "I want to know if women are waiting longer now to get married than they did 10 years ago" is precise, simple, and straightforward. We know we are looking at women. We know we are going to compare average age of marriage now with average age of marriage for women 10 years ago, thus setting the time parameters and the variable under study (age).

The research question "I want to know if women are waiting longer now to get married than they did 10 years ago, and I want to know if that wait, if present, is related to education and/or career choices and varies by ethnicity" is much more complicated. The second part of the question is dependent on findings from the first part: If women are not waiting longer, the issues of education, career, and ethnicity become potentially moot. This second research question is not very precise.

Research questions should be stated in a way that does not presuppose any particular outcome—that is, stated in the most **objective** manner possible. For example, asking "Is there a difference in intelligence between men and women?" is much more objective than saying "Men will be found to be more intelligent than women." The second research question presupposes the outcome; it is not really a question as much as a statement of anticipated findings based on stereotypic assumptions.

As a slight aside, there are instances where researchers may have slanted data to fit preconceived notions of findings. Samuel G. Morton, a physician living in the 1800s, conducted research on craniometrics. He deduced that there were variations in intelligence by race based on cranial volume. Stephen J. Gould has suggested that Morton modified his measuring technique to achieve a preconceived notion of racial intelligences (Gould, 1981).

Research questions should also lend themselves to **operationalization**. Operationalizing a research question means that you ask the question in such a way that the question clearly defines what variables will be compared and that the variables under investigation are measurable. Some constructs in psychology are difficult to operationalize. One has only to think of Freudian theory and realize that it might be very difficult to find or construct a measure of the id or ego, much less the concepts of Eros (life) or Thanatos (death).

Finally, research questions should be **rational**. "Well, that's obvious," you say. It certainly should be, but one of your authors is reminded of a student from a number of years ago who was in a thesis class. The purpose of the class was to get students focused on a research question that was workable and started on the initial chapters of their theses. The student, when asked what he wanted to study, replied, "God." The two professors conducting the class, along with other class members, attempted to clarify what the student meant. Did he mean religion or spiritual beliefs, both of which can be studied? "No," he replied, "I want to study God." "Well, exactly what aspect of God do you want to study? Manifestations reported in the Bible during the pre-Christian era?"

No matter how we probed over the next few weeks, the student adamantly stated that he wanted to study God. In desperation, we took another tack and asked how he was going to study God methodologically. He suggested factor analysis. My colleague looked at the student somewhat in disbelief. Then, in frustration and desperation, blurted out with a loud laugh, "You mean, like, you are going to factor-analyze *God*?"

This was not a doable thesis project. It was somewhat ludicrous, actually. We were never able to get past this rather irrational plan to study God. Eventually, the student dropped out of the clinical psychology program and entered a program that trained religious counselors—probably a much better fit.

Apart from studying God, some students, in carving out research areas for theses or dissertations, simply take on too much, adding a degree of irrationality in terms of what they can hope to do in one study. Another student of one of your authors had a research idea for his thesis that would have taken, at best, 10 to 15 years to complete. It was complex and elaborate, essentially seeking to distill a large field of research down into a compact theory based on a modicum of data. It was an impossible task. The student, however, insisted on pursuing this despite all cautions to the contrary. The end result of this rather irrational insistence on solving a whole field of research in one study was that the study was never completed. Irrationality does come in different forms.

Stating the Research Question

While research questions deal with the what, how and why, and when and where, they must be phrased in a way that leads to a testable investigation. From the research question, the researchers develop predictions, or educated guesses—known as hypotheses, about the expected outcomes of the study. A **hypothesis** is a statement or proposition about the characteristics or appearance of variables or the relationship between variables that acts as a working template for a particular research study.

For example, "It is hypothesized that variable X is related to variable Y" is a hypothesis that would direct us to gather data on the two variables and then examine those data for a relationship. This would be a descriptive hypothesis.

If we said, "Variable X is a primary influence on the occurrence of variable Y," then we would gather data on the two variables and look for a causal relationship. This would be an inferential hypothesis.

It used to be that hypotheses were always stated in the null format. Had any of your authors stated the hypotheses for our dissertations in any other format, we probably would have been marched out of the academic building and placed in a public stockade. The **null hypothesis** (symbolized by H_o) is essentially a statement of no difference.

Let's go back to our earlier example of a descriptive hypothesis, "Women waiting longer now to get married than they did 10 years ago." As a null hypothesis, one would say, "There will be no difference in the average age of women who get married today as compared to 10 years ago."

To many students, this makes no sense. If you think women are waiting longer now, and you think this is worth investigating, then why say they are not waiting longer—that there will be no difference in their average ages? You are essentially stating that there is no difference now as compared to 10 years ago, yet you believe there is a difference, which is why you are doing the research—and that makes no sense. The null hypothesis is really a statement of the similarity of two random groups before any manipulation is conducted.

Suppose we had two random samples from the same population. They should mimic each other, given the assumptions of random sampling. If you then say there is no difference, you are describing those two random samples at the point they were selected—and there should be no difference. The question now is whether your manipulation or treatment will make them different, in which case, you will reject the null hypothesis of no difference.

A quick example will illustrate this concept. Suppose you want to study the effects of two medications on depression. You select two random groups from a large population of depressed individuals. Members of those two groups should mimic each other in gender, age, weight, height, etc., and degree of depression because of random sampling. You administer one drug to one group, the other drug to the other group. After three months, you measure levels of depression using the Beck Depression Inventory. If there is now a difference, you will reject the null hypothesis of no difference.

Fortunately, for students, the **alternative hypothesis** (symbolized by H_a) is becoming much more acceptable. The alternative hypothesis states the research question directly as follows: "Are women waiting longer now to get married as compared to 10 years ago?" This is straightforward, and collected data will either say yes, they are, or no, they are not.

In current research studies, you may encounter the hypotheses examined in either of these ways: traditionally, by assuming the null condition, or more progressively as a direct examination of the alternative condition.

Sources of Research Questions

Where do research questions come from? Go back to Section 1.1 and the chapter on "Why do research?" We could certainly say that research questions come from curiosity and self-protection. At a more pragmatic level, research questions are generally derived from existing **theories**, **models**, **constructs**, or **replications**.

A *theory* is a statement of the inter-relationship of variables, behaviors, or events that purports to explain some phenomenon in detail. Theories generally grow out of *models*—being less-developed and -integrated statements that purport to explain some phenomenon. With sufficient time and research, models may mature into theories.

Constructs are words that are used to describe something we think exists based on behavior, observations, or investigation. For example, we might surmise that a couple walking down the street hand-in-hand holds some degree of affection for one another, yet affection is not something that exists in a physical sense—we cannot dissect a person and find a physiological site for "affection." It is an inferred condition based on our observation. Other constructs in psychology would include intelligence, love, hate, anger, depression, psychosis, etc. When you think about it, the study of human behavior is replete with constructs.

A **replication** of some existing study or an expansion of an existing study could be carried out in an attempt to match the initial study exactly, or could duplicate the methodology, but with a variation in the characteristics of the subjects. Much research (particularly theses and dissertations) concludes with a section that suggests directions for further research for anyone wishing to investigate what was presented. Such directions for future research pave the way for an expansion of existing research.

As mentioned earlier, research generally moves in small steps where individuals take existing research and build upon it. Such a process would certainly include replicating studies already conducted to clarify or verify results from those studies, as well as expanding on existing research.

Where to Next?

Next, in Chapter 2, we will look at the process of research. We have already established that it is a step-by-step process, but will look more specifically at the perpetuation of research and building the literature base to support the research question and research method, along with some advanced organizers and factors that threaten research. There is more—much more—to talk about.

Key Terms

- Alternative hypothesis
- Constructs
- Dependent variable
- Design
- Experimental design
- Hypothesis
- Independent variable
- Inferential statistical techniques
- Method

- Methodology
- Model
- Null hypothesis
- Operationalization
- Parsimonious
- Piltdown man
- Precise
- Qualitative design
- Quasi-experimental design

- Rational
- Research
- Research question
- Replication
- Scientific method
- Single-subject design
- Statistics
- Survey design
- Theory

Questions

1. Why is a research course important?

2. Why do humans conduct research?

3. What are some potential sources of false beliefs?

4. What is the difference between descriptive and inferential statistical techniques?

5. Explain why the null hypothesis is a statement of no difference.

6. Which characteristic of a research question do you think is most important and why?

7. What are some potential sources of research questions?

8. Write a null hypothesis and then write the alternative hypothesis for the null hypothesis.

9. Operationalize the following research question: Taller men earn more per hour than shorter men do for the same work.

10. What are the basic tenets of the Scientific Method?

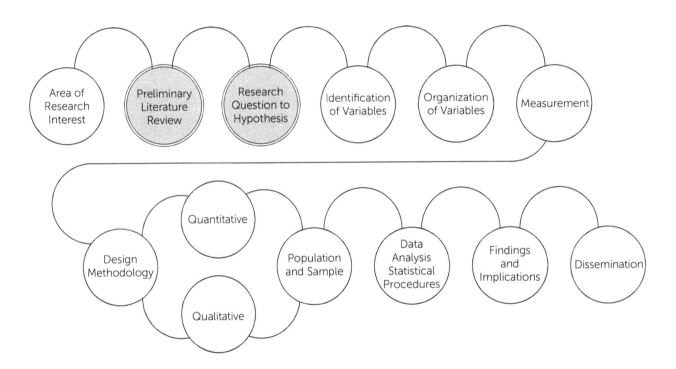

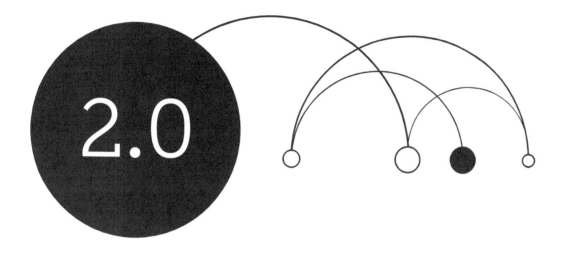

2.0

Research as a
Step-by-step Process

It has been proven by many research studies that celebrating birthdays is healthy. The data show that those who celebrate the most birthdays become the oldest.

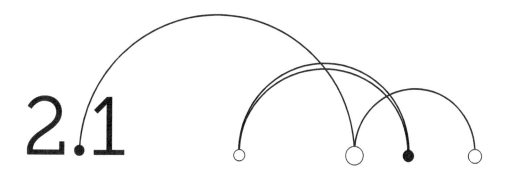

2.1

How research self-perpetuates: The process of building on itself

Research is a process, meaning that it is a series of steps taken in progression to arrive at an end or goal state, such as the answer to a question or resolution to a problem. As part of this process, research is also iterative, which means that each step in the process builds on the previous steps. In the pursuit of scientific knowledge, this is critical. Humans and their technology have progressed based upon this process of accumulating knowledge, applying that knowledge and taking the next steps forward in pursuit of new knowledge. The top of one mountain is the bottom of the next, so to speak.

In this way, research is a dynamic and never-ending process in which we, as researchers, try to build on and extend the work of others and the state of the field in general. From all this, new ideas emerge, older ideas are modified, and collectively we move forward in understanding.

In the conduct of actual research, this building on the existing knowledge base drives the process at all stages. It helps define the development of appropriate research questions by identifying what has been asked before and how it has been asked and answered. It outlines the theoretical framework into which we will merge our research. It offers parameters for the consideration of variables, populations, samples, types of measurements, methodologies, and analyses. In conducting research, when we "stand on the shoulders of giants," we become a collaborative partner in the larger quest for knowledge, whether for the sake of knowledge or in the pursuit of a practical solution to a problem.

The **literature review** and its associated **annotated bibliography** represent the ways in which a researcher identifies what is already known about the general topic of interest. The process of preparing an annotated bibliography and literature review will be explored in depth in the next sections.

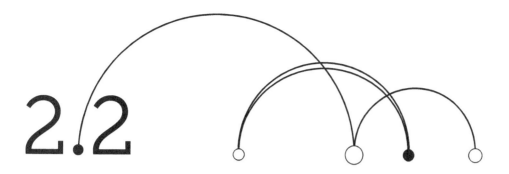

2.2

How to conduct a literature review based on a research question

Most research begins with a general statement of the problem—or rather, the purpose for engaging in the study. From this, a testable, or explorable, research question is developed—one that will form the foundation of the entire investigation. In quantitative studies, the research question explores the relationship between two or more identified variables. In qualitative studies, the research inquires about more subjective aspects of a process or phenomenon. Research questions provide the basic structure and focus for literature reviews.

Before details of the research question can be finalized and you can get on with the rest of your study, it is important to conduct a thorough review of the literature to understand what is known and what remains unknown about a topic, issue, or problem. *Literature*, in this context, refers to the complete body of knowledge that exists about a particular subject. The literature review will summarize the status of existing research related to your general topic. You will interpret the findings of previous research and identify contradictions, controversies, and gaps in the knowledge base.

Specifically, the literature review will explore some of these questions:

- Has your research question been studied before? If so, to what extent?
- What were the results? Is there agreement or disagreement among the existing studies?
- What variables and measures were used to assess the constructs of the study? Were they appropriate? What alternatives exist?
- How were data collected and analyzed?
- What methodologies were used to investigate the topic? Were they sound and appropriate?

- What populations and samples were used in existing studies?
- What conclusions, recommendations, and areas of further study were suggested?

The literature review helps ensure that you do not replicate an existing study or set of studies—that is, unless such replication is the point of your research, in which case, those studies would serve as a guide for decisions about what aspects to keep constant and what aspects to change.

Generally, the literature review identifies flaws or holes in previous research, providing justification for your study. A gap in the current research, as identified by a thorough review of the literature, solidifies the direction for new research questions.

Remember that the process of research is iterative, involving repetition that fine-tunes the researcher's course of action toward the final goal of adding to the field of knowledge. The final development of a well-honed and focused research question is the product of refining the original question by surveying the existing literature. In their book on the technique of writing effective literature reviews, Jesson, Matheson, and Lacy (2011, p. 20) outline the steps of refining of the research question this way:

1. Formulate a draft research question.
2. Search for information.
3. Skim, scan, read, reflect, and search some more, defining key concepts.
4. Obtain articles and read some more.
5. Reassess the question.
6. Formulate the final research question.

Using this process, your research question will flow logically and naturally from what has already been discovered and what still has to be discovered about a topic.

Note that you will also refer back to the existing studies in your literature review when you discuss the results of your own study. After you have analyzed your data and determined conclusions based on that analysis, you will circle back to explain how your results compare to those obtained by other studies. In this way, your new research adds to the field of knowledge in a particular area and becomes part of the existing literature that will be referenced in future studies. This iterative process is how the overall process of research continues.

A word about negative results. As you probably already know, what we typically set out to do when we conduct a research study is to show how the data support our hypotheses. We do this objectively, honestly, and without bias, of course—but since the hypotheses represent our best guesses at the relationships between the underlying phenomena, we expect that this is the truth that our data will reveal.

However, sometimes it doesn't work out that way ("The best laid plans ...."). Sometimes our results fail to confirm our well-grounded suspicions. Sometimes (actually, many times) the results are the opposite of what we expect, or worse yet, inconclusive. This state of affairs is sometimes referred to as "negative results."

If you think about the purpose of research and the fact that individual studies and projects are developed based on the pre-existing knowledge and results of other studies, it becomes easy to see that these negative results are as important as, if not arguably more important than, results that are not surprising and confirm what we already thought we knew. Given the long and often arduous process of research, we arrive at hypotheses only after a critical examination of the existing literature on the topic and a logical extrapolation of that knowledge onto the conditions of the current study.

Researchers put a lot of time and logical thought into the various steps of the process to ensure that there is a sound premise for the hypotheses and the details of the methodology. If the study yields unexpected results, it should be a red flag for the researchers of that study, as well as for others in the field, that something, somewhere, in the process, has to be re-examined and reassessed. Certainly, there are numerous reasons why a study may not produce the results we expect, but each of those reasons is a topic for consideration as the research moves forward.

Certainly, these negative results should be shared among professionals in the discipline, since they represent important insights, either into the phenomena being studied or the way the research was conducted. However, in the current state of research in most fields, there is an underlying bias against publishing negative results. It's not an official policy, but more a *de facto* state of affairs. In fact, many researchers have spoken out against this trend—the 2017 revised European Code of Conduct for Research Integrity has mandated that academics and journals treat negative experimental results as being equally worthy of publication as positive ones. However, the unofficial trend not to publish inconclusive or conflicting findings is still pervasive. Perhaps it would do us well to remember the words of Mark Twain: "What gets us into trouble is not what we don't know, it's what we know for sure that just ain't so."

The annotated bibliography is a useful tool for organizing the various sources related to your topic. The next section outlines how to construct an effective annotated bibliography.

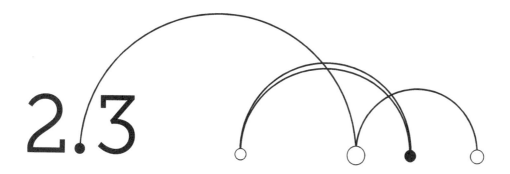

2.3

Developing an annotated bibliography

A bibliography is a list of references (books, journals, websites, periodicals, etc.) that the researcher used in investigating a topic. It typically lists pertinent identifying information about the source (author, title, publication date, publisher, etc.) organized in a standardized format (e.g., APA, MLA, Chicago, Turabian). An annotation refers to a summary and/or evaluation, so an annotated bibliography includes both of these components for each of the listed sources.

The Purdue University Online Writing Lab (OWL) recommends addressing the following questions in an annotation:

- **Summarize**
    - What are the main arguments?
    - What is the point of this book or article?
    - What topics are covered?
    - If someone asked what this article/book is about, what would you say?
- **Assess**
    - Is it a useful source?
    - How does it compare with other sources in your bibliography?
    - Is the information reliable?
    - Is this source biased or objective?
    - What is the goal of this source?
- **Reflect**
    - How does it fit into your research?
    - Was this source helpful to you?
    - How does it help you shape your argument?
    - How can you use this source in your research project?
    - Has it changed how you think about your topic?

The annotated bibliography will be useful in constructing the literature review. Typically, the sources you will use in the literature review will be a subset of those in the annotated bibliography. The annotated bibliography will provide a reference for those sources that will be most helpful. During the course of constructing your literature review, you will probably come across additional sources that are relevant, but the annotated bibliography gives you a good place to start.

In constructing an annotated bibliography, you should include a variety of sources, such as books; academic, scientific, and trade journals; periodicals; government documents; and various websites. Your topic will determine the relative relevance and value of the numerous types of sources, as well as the importance of the age of the research. Topics related to state-of-the-art technology and other time-sensitive issues will obviously have to rely more heavily on current research than topics with a more historical perspective. However, even for studies that deal mainly with current phenomena, some level of historical grounding may be appropriate.

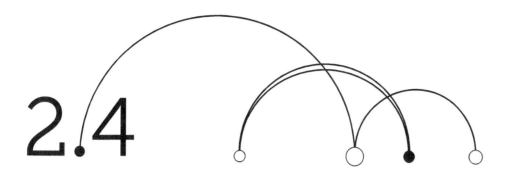

2.4

Organizing a literature review

Depending on the context for which it is written, a literature review can vary somewhat in format, content, and organization. A typical literature review comprises:

- Introduction
- Main body
- Conclusion/Summary
- References

The introduction is where you present the general topic and define the scope of the inquiry. This may include a description of the exclusions or limitations of the subject to be considered, especially if the topic is very broad or the area under investigation is very specifically focused. A description of the range of sources and/or current state of research also can be included in this opening section. The new research question can be introduced here as well.

The main body of the literature review is where the historical background and existing research are summarized, explained, critiqued, evaluated, and synthesized. This section also provides a description of underlying theories and constructs that support the various studies, as well as the practical problems that warrant exploration of the topic. In addition, relevant terminology and jargon specific to the subject are explained, as needed, for a clear understanding of their use in this context. Depending on the topic and the nature of the available research, this section can be organized in various ways, including categorically, chronologically, or thematically. Relevant section headings should be used for clarification.

The conclusion/summary section reviews the major findings and puts the new research question in context. It provides an opportunity to justify the significance and purpose of the study being proposed. Specifically, this is where you highlight how the knowledge base contributes to the field and explain how the new research addresses a gap in the existing literature.

The references section is a critical component of the literature review. It differs from the list in the annotated bibliography because it includes only those sources specifically cited in the review and does not include the annotation. Strict adherence to the formatting protocol is important throughout the paper, but especially in the reference list, so others can get access to the associated studies.

The American Psychological Association (APA) publishes style guidelines for formatting different types of papers, documenting sources, and adhering to standards of clarity and mechanics. Their current guidelines are published in *The Publication Manual of the American Psychological Association*, 6th edition. The next section outlines the major points of citing sources in APA format.

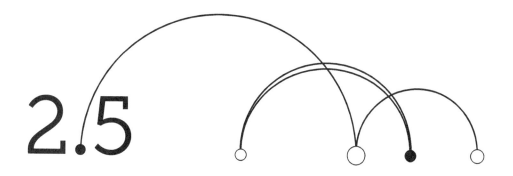

2.5

APA publication manual guidelines

Citations and Why You Must Cite!

Oh, the dread, and what a pain. "Really, do I have to provide citations in my paper?" The answer of course is *yes*, and this section will help you understand why, as well as how to provide citations and avoid plagiarism.

Citations must be used because they are required in academic writing, but, more importantly, because using them is just good research practice, as well as ethical procedure. Providing citations for works you have consulted in developing your research shows that you knew how to engage in the process of research and that you respect the contributions of others in your field. You do not want your work to be considered **plagiarism**!

Plagiarism—using someone else's work as if it's your own by including it without citing the original creator of the work—can lead not only to academic suspension or dismissal but to losing your job. Consider Karl-Theodor zu Guttenburg. Guttenburg became the German minister of defense at age 39 and tipped to be the future chancellor of Germany. He came from a noble background going back to 1315. However, in 2011, it was discovered that he had plagiarized part of his dissertation. Within two weeks of the discovery, the University of Bayreuth stripped him of his PhD, and he resigned in shame, retreating to his family castle in Bavaria, never to be heard from again.

In 2013, after the demise of Guttenburg's career, another German government official, Annette Schavan, quit over plagiarism charges and was stripped of her PhD—a somewhat ironic situation, since she was the minister of education! Now, most of us will never be on the cusp of greatness or reach such levels of importance, but we never know where life will lead us.

There are several citation systems around, such as Modern Language Association system (MLA), which is typically used more in citations for the liberal arts and humanities. Others include the *Chicago Manual of Style* (used widely in publishing); *American Medical Association (AMA) Manual of Style* (used by many medical journals); and those of the Modern Humanities Research Association (MHRA); Vancouver, Columbia, Oxford, and Harvard systems; American Sociological Association (ASA); and Institute of Electrical and Electronics Engineering (IEEE) style, among several others. Each system has its advantages and disadvantages, but we will keep it easy by sticking with APA style for the rest of this discussion.

APA is used mostly in the social sciences, and many journals and universities require APA for citing papers. However, these citation systems cover more than just how to cite a reference; they outline how things such as gender should be addressed, how charts should be displayed, and what style of writing should be used.

A citation is a way—a format—to reference a source that is typically embedded in the body of the work, with more detail provided in the reference section at the end of the work.

But Why Cite?

There are three main reasons to provide citations in your paper.

First, show that you have done your research. Providing appropriate citations indicates to the reader that you have explored the relevant literature on the subject comprehensibly, which gives relevance and weight to your research.

Second, properly credit work and ideas to the correct sources. The reader needs to know where ideas have their genesis: with you—the writer—or with another author. This leads to intellectual honesty and helps avoid plagiarism. *Plagiarism* is using someone else's work (such as ideas, findings, or words) without providing credit where due, making it seem as if you wrote something that you did not. Yes, you can take someone's thoughts and summarize them in your paper, but make sure to provide a citation, and then you are good to go. If you want to use their direct words, you must cite them properly, using a direct quote (see examples below). Please note, though, that direct quotes should be used sparingly in your writings.

To avoid plagiarism, when you are drafting something, read the literature, form your own ideas, put the paper away, and then write about it using your own words. When providing an idea that is close to someone else's, simply insert an in-text citation—usually author's last name and date the material was published. If you write this way, you will not have to worry about plagiarism. However, if you do not and your work is found to have been plagiarized, then understand that you can face a heavy penalty, including expulsion from a university. Many universities use text-matching services that will identify copied or slightly changed text. Check out www.plagiarism.org for more information.

And remember that just because you find information on the Internet, you can't repeat it in your research paper without crediting the original source. Many people nowadays—not only students, but even adults—think that information on the Internet is fair game to reuse as is without citations. That is simply not the case, and not ethical.

Finally, providing citations allows the reader to find and independently review the author's arguments and check the sources in case they have any questions. Citations provide a "bread crumb" trail for a researcher to follow to work back and review the original source. One of your authors, many years ago, was reading an interesting student paper and stumbled onto a statement that a student had made that provided part of the answer to a question that your author had been pondering for a while. Luckily, the student had provided a citation to the source document, which in the end helped provide the answer to that question. Just imagine: A simple student paper helped this professor advance his own research.

Peer-reviewed literature

We would like to take a moment to talk about peer-reviewed journals and the importance of using them in your academic writing.

First, why is this important? If you are reading this book, then you are most likely a university student somewhere in the world and, as such, are learning to synthesize information, not just regurgitate it. To do so, you must first present an argument and then back it up with something that has already been well established. Anyone can say whatever they like, but it does not mean what they say is true. Using peer-reviewed research brings more credibility to what you are trying to argue.

Second, what is peer-reviewed research (sometimes also known as *refereed research*)? Peer-reviewed research has been examined and scrutinized by experts in the field of study to ensure that the research was properly conducted and the findings are accurate or reflect sound methodology (at least at that point in time using that methodology).

Third, how do you access peer-reviewed research (the important part!)? Ninety-nine percent of the material on the Internet is *not* peer-reviewed. It may look like it, but it usually is not. Even a lot of the material on Google Scholar is *not* peer-reviewed. However, this is changing, and some of the more recent open source material that can be found on the Internet may have been peer-reviewed. The best place to find (and ensure) peer-reviewed articles is the library. We hope that by this point in your academic career, you know how to access the library at your university—mostly online nowadays. Whatever you do, please do not use Wikipedia.

One last suggestion: In general, it is best to not use old articles, since new knowledge may have come to light since they were published, and the older findings may no longer be valid. Try to stick with articles that have been published in the last three to five years. But beware: Some articles are considered seminal in their fields, such as Latane and Darley on bystander behavior or Altman (1968) on corporate bankruptcy.

APA

The Publication Manual of the American Psychological Association (APA) is the guide to understanding how to format your paper properly, provide in-text citations, and build the reference table at the end of your paper. We will provide some general guidelines on APA formatting, but for more information, you should reference the manual. Just to be clear (and to avoid plagiarism): Some of the examples in the next section were obtained directly from the APA style manual. For the sake of clarity and space, and since these all are based on APA, page-number citations have been omitted.

How references are constructed

APA references have four general elements:

- Names of the author(s)
- Date of the publication
- Title of the work, and
- Source of the data.

Therefore, in general, an APA reference follows the following format:

> Author, A. A. (year). Title. Source.

For a little more detail, here is the general reference format:

> Author, A. A., Author, B. B., & Author, C. C. (year). Title of article. Title of Periodical, x(x), pp–pp. http://dx.doi.org/xxxxxxx.

Note that you do not use the first or middle names of the authors—only initials. This is to reduce gender bias in academia.

An example of the APA reference for a journal article published by some of your authors is:

- Author: Walton, R. O. & Politano, P. M.
- Year: 2016
- Title: Characteristics of general aviation accidents involving male and female pilots
- Source: *Aviation Psychology and Applied Human Factors*
- Vol: 6
- Issue: 1
- Pages: 39–44
- doi #:10.1027/2192-0923/a000085

The proper APA reference for this information would look like this:

> Walton, R. O., & Politano, P. M. (2016). Characteristics of general aviation accidents involving male and female pilots. *Aviation Psychology and Applied Human Factors, 6*(1), 39–44. http://dx.doi.org/10.1027/2192-0923/a000085

Figure 2.1 provides the same citation, but with a little more detail.

FIGURE 2.1 Sample APA Reference With Notes

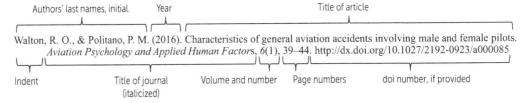

The reference for a book would look like this (note that the title of the book is italicized):

Walton, R. O. (2015). *Predicting financial distress in the all-cargo airline industry.*
Saarbrücken, Germany: Lambert Publishing.

If no date is available then you can use "n.d." (for no date) in your references.

How to do in-text citations

In the last section, we covered how to build your reference table, but we skipped ahead a little. Before you build a reference table, you have to find something to cite and somehow make a notation in the actual text. To do this, you simply insert an in-text citation. APA uses an author-date system (author, date). For example, the paper referenced above that was written by Walton and Politano could be cited in one of several ways.

For a direct quote you could state:

Walton and Politano (2016) stated that Amelia Earhart "ground looped her Lockheed Electra on take-off on March 20, 1937, at Ford Island field in Hawaii" (p. 39).

Or you could have a direct quote with the author and date, along with the page number at the end of the quote, such as:

Amelia Earhart "ground looped her Lockheed Electra on take-off on March 20, 1937, at Ford Island field in Hawaii" (Walton & Politano, 2016, p. 39).

If you are not using a direct quote, then you only need to provide the author and date for the citation, such as:

In 1937, Amelia Earhart crashed her aircraft during take-off in Hawaii (Walton & Politano, 2016).

Note that when you enclose the author's name in parentheses, you use the ampersand (&) sign between the authors' names, while when you state the authors openly in the text (such as in the first example), you use "and" between their names. A period goes at the very end of the citation. Finally, make sure that for every in-text citation, you have an entry in the list of references that corresponds to the citation.

Here are some examples of actual citations in the Walton and Politano (2016) paper referenced above:

This study extends and updates the research by Baker, et al. (2001) and Bazargan and Guzhva (2011) on the characteristics of GA accidents based on gender.

Bazargan and Guzhva (2011) examined accidents between 1983 and 2002 found in the NTSB database at that time.

Vail and Ekman (1986) stated that "perhaps for greater safety, there should be two pilots in every cockpit, a male for taxiing and a female for every other phase of operation" (p. 302).

GA accident studies as a whole have one flaw, in that the majority of studies "aggregate all 14 CFR Part 91 operations inclusive of pilots holding various licenses as well as trainees" (Shao, Guindani, & Boyd, 2014, p. 371).

And, finally, when to use "*et al.*" In most cases, you can use the phrase after the first time you fully cite a source. For example, the second time that Walton and Politano (2016) cited Shao, Guindani, & Boyd in the paragraph below, we changed the citation to read (Shao, et al., 2014, p. 371):

GA accident studies as a whole have one flaw, in that the majority of studies "aggregate all 14 CFR Part 91 operations inclusive of pilots holding various licenses as well as trainees" (**Shao, Guindani, & Boyd**, 2014, p. 371). Similarly, in many studies, "accidents for single and multiple engines are grouped despite the fact that the latter carry an increased risk of fatality" (**Shao, et al.**, 2014, p. 371). [*Note:* bold added for emphasis].

More examples

To provide you some examples, here is an extract of a paper by Walton and Politano published in 2014. The proper APA reference for this paper is:

Walton, R. O., & Politano, P. M. (2014). Gender related perceptions and stress, anxiety, and depression on the flight deck. *Aviation Psychology and Applied Human Factors*, *4*(2), 67-73. doi:10.1027/2192-0923/a000058

Pay careful attention to how we provided the in-text citations to attribute the facts stated to the proper authors. Please note that manuscripts are typically double-spaced and span the whole page with 1-inch margins; however, in the final published paper, the publisher will often place the text in two single-spaced columns, as in this example:

Gender perceptions are attitudes held by a society that influence how men and women growing up in that society are inculcated into the gender system (Liedberg, Björk, & Hensing, 2010). These attitudes are a function of the history, culture, daily interactions, and social norms of the society and heavily influence gender stereotypes and roles, beliefs about strengths and weaknesses of the genders, occupational choices open to the genders, choice of college majors, etc. (Bradley, 2000; Eagly, 1987; Li, 1999; Liedberg et al., 2010). Within these gender perceptions, there are few occupations considered to be gender-neutral (Diamond & Whiteside, 2007; Lippa, 2005). The masculinization or feminization of an occupation often defines who enters the occupation, thus producing at some level a collusion by the genders on the continuation of gender-specific perceptions of particular occupations (Atkinson, 2009).

Entry by women into male gender–specific occupations often results in sexism, high scrutiny, isolation and ostracism, less favorable advancement opportunities, harassment to include sexual harassment, gender-role stereotypes, work–home conflicts, higher occupational dropout rates, and the need to adapt by assuming gender-incongruent behaviors (Cohen & Huffman, 2003; DiDonato & Strough, 2013; Germain, Herzog, & Hamilton, 2012; Huppatz & Goodwin, 2013; Watson, Goh, & Sawang, 2011). Consistent with role strain theory (Kantor, 1977), these are all attitudes which, if held in the workplace, are negative in nature and can produce stress, strain, and anxiety for the targeted

individuals (Cohen & Huffman, 2003; DiDonato & Strough, 2013; Huppatz & Goodwin, 2013). This is especially true for women, who are often devalued anyway (Germain et al., 2012). The sequelae to workplace stress may include psychological and physiological reactions such as, but not necessarily limited to, emotional distress such as anxiety and depression, frequent headaches, muscle tension, change in sleep pattern or change in appetite, digestive problems, hormonal changes, muscular system changes, increase in heart rate, and changes in the immune and metabolic systems (Rice, 2012). These manifestations may be more evident in women when the work environment is seen as hostile or as a threat (Watson et al., 2011), particularly where interpersonal conflict, the primary source of occupational stress, may be high (Mazzola, Schonfeld, & Spector, 2011).

Research has consistently shown that women in such gender-incongruent roles are significantly more stressed than their male counterparts; experience greater degrees of health problems; report lower levels of self-efficacy, self-esteem, and self-worth; experience a devaluing of their contributions; hold lowered expectations of success; and perceive lower support and commitment from the organization in comparison with their male counterparts (Germain et al., 2012; Heilman & Okimoto, 2007; Heilman, Wallen, Fuchs, & Tamkins, 2004; Jacobs, Tytherleigh, Webb, & Cooper, 2010). By contrast, men working in female-specific occupations may not experience the same negative

Tables in APA

Tables formatted for APA papers are relatively easy to create but are often improperly constructed by students. Table 2 shows how to set up a table properly for use in academic papers. Note that there are only lines at the top and bottom of the table, and a line under the title line. Other than those, there are no lines, either vertical or additional horizontal ones). Also note that the table starts with the words "Table 2," then two spaces, the title of the table (italicized), and then the table. Tables should all be in Times New Roman, 12 point, and can be single- or double-spaced. If you use a note at the end of the table, start with the word "*Note*" in italics and insert the note (such as where you obtained the data).

Table 2

Gender Differences for NTSB Landing Phases

Landing phase	Percent males	Percent females	Z	p
Landing	16.22	9.35	1.89	.0588
Flare/touchdown	40.57	53.27	-2.587	.0097*
Landing roll	43.21	37.38	1.181	.2376

Note. * Significant differences

General paper setup in APA

In general, an APA manuscript should be set up using the following format:

- All text double-spaced (exceptions can be made for large tables)
- Times New Roman typeface, 12 point
- Running header at the top of each page (upper left-hand side)
- Page number in the upper right-hand corner
- 1-inch margins
- Each paragraph indented (except for the abstract)

Papers should have the following basic layout:

- Title page
- Abstract
- Index (normally not needed for shorter papers)
- Content of your paper (each title below should be bold and centered)
    - Introduction
    - Literature review
    - Methodology
    - Results
    - Discussions and conclusions
- Reference table

Figure 2.2 provides an example of a properly formatted APA manuscript.

FIGURE 2.2 Properly Formatted APA Manuscript

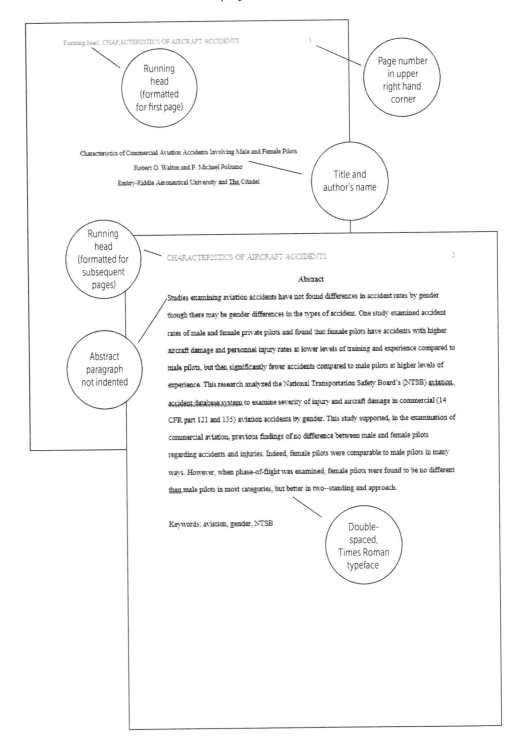

CHARACTERISTICS OF AIRCRAFT ACCIDENTS   3

Introduction

The accident rate of commercial aviation operators in the United States is very low, however accidents and mishaps do occur from time to time within the industry. Research within governments, industry, and academia have continued to reduce aviation accidents to a point that a commercial aviation accident is truly a rare event. There are very few women in aviation, but this number is slowly growing. As the number of female pilots increases so will the need to understand gender related issues that could affect flight safety. Because of the disproportionately small number of women in aviation, there have only been a few commercial aviation accidents involving a mixed gender crew. One such example is Colgan Air flight 3407, which had a male captain, and a fem

Transpo

where

comm

accident, p

environment and a

U.S. Comm

of Federal Regulat

common carriers i

pilot for a Part 121

certificate. Holder

and allows a pilot

obtain an Airline T

pilot must have log

1-inch margins

Title centered and boldface

Table number

CHARACTERISTICS OF AIRCRAFT ACCIDENTS   9

At Experience Category V (5000 hours and up), there were no significant gender differences in pilots related to aircraft damage (See Table 1). There was a significant difference in male and female pilots in terms of s[...]ith females higher (30.65%) than male pilots (18.93%), z=-2.331, p=.019[...]ury category where there was a significant difference.

Table title italicized

Lines top and bottom and after heading

Table 1

Experience Categories (CAT I, II, III, IV, V) by Damage Levels for Male and Fem[...]
Expressed as Percentages

FAR Part 61 Categories	CAT I Male/Female	CAT II Male/Female	CAT III Male/Female	CAT IV Male/Female	CAT V Male/Female
No Damage	n/a	n/a	(low n)	13.92/15.91	25.31/33.87
Minor Damage	n/a	n/a	(low n)	14.07/10.23	23.04/27.42
Substantial Damage	n/a	n/a	59.21/62.16	53.44/45.45	41.33/30.65
Aircraft Destroyed	n/a	n/a	20.79/27.03	18.56/28.41*	10.32/8.06

*Significance of the difference in proportion, z >1.96, p <.05
Note: Commercial pilots tend not to fall into CAT I or CAT II due to minimum flight hour requirements.

Additional analysis compared male and female pilots on accidents and injuries by phase-of-flight. No significant differences were found for taxing, takeoff, climb, in-air maneuvering, descent, or landing. Female commercial pilots (2.05%) however, had significantly fewer accidents than male pilots (7.31%) involving damage to the aircraft and/or injuries while standing on the ramp, z=2.43, p=.0157. Such accidents might involve not having the aircraft properly positioned, accidental movements, etc. Female pilots (3.42%) also had significantly

"Note:" italicized

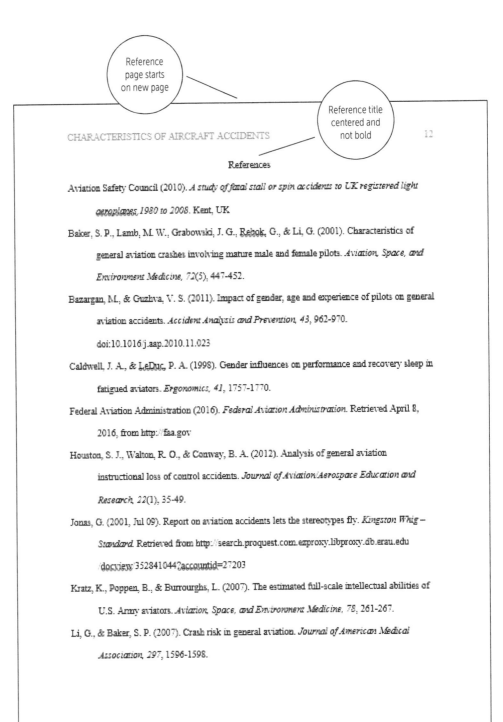

Reference page starts on new page

Reference title centered and not bold

CHARACTERISTICS OF AIRCRAFT ACCIDENTS 12

References

Aviation Safety Council (2010). *A study of fatal stall or spin accidents to UK registered light aeroplanes, 1980 to 2008.* Kent, UK

Baker, S. P., Lamb, M. W., Grabowski, J. G., Rebok, G., & Li, G. (2001). Characteristics of general aviation crashes involving mature male and female pilots. *Aviation, Space, and Environment Medicine, 72*(5), 447-452.

Bazargan, M., & Guzhva, V. S. (2011). Impact of gender, age and experience of pilots on general aviation accidents. *Accident Analysis and Prevention, 43*, 962-970. doi:10.1016/j.aap.2010.11.023

Caldwell, J. A., & LeDuc, P. A. (1998). Gender influences on performance and recovery sleep in fatigued aviators. *Ergonomics, 41*, 1757-1770.

Federal Aviation Administration (2016). *Federal Aviation Administration.* Retrieved April 8, 2016, from http://faa.gov

Houston, S. J., Walton, R. O., & Conway, B. A. (2012). Analysis of general aviation instructional loss of control accidents. *Journal of Aviation/Aerospace Education and Research, 22*(1), 35-49.

Jonas, G. (2001, Jul 09). Report on aviation accidents lets the stereotypes fly. *Kingston Whig – Standard.* Retrieved from http://search.proquest.com.ezproxy.libproxy.db.erau.edu/docview/3528410442accountid=27203

Kratz, K., Poppen, B., & Burrourghs, L. (2007). The estimated full-scale intellectual abilities of U.S. Army aviators. *Aviation, Space, and Environment Medicine, 78*, 261-267.

Li, G., & Baker, S. P. (2007). Crash risk in general aviation. *Journal of American Medical Association, 297*, 1596-1598.

Summary

The APA manual is well more than 200 pages long, so there is no way that we can summarize the entire manual in this book. However, we have provided the basic elements to allow you to reference material properly and avoid plagiarism. Use these examples as a starting point, but make sure to review the APA manual for more details on not only proper referencing but also on how to improve your academic writing for clarity, punctuation, and other mechanics of writing.

Where to Next?

In the next chapter, we will explore the theoretical framework of research. We will take a deeper look at research questions, the types of variables involved, and their purpose in the research process.

Key Terms

- Annotation
- APA
- Bibliography

- Citation
- Literature review
- Peer review

- Plagiarism
- Reference

Questions

1. What is the purpose of a literature review?

2. What is the difference between an annotated bibliography and a literature review?

3. How does a literature review help shape research questions?

4. What is the purpose of peer review?

5. What is the purpose of a standardized formatting convention such as APA?

6. What are four reasons you should provide citations in your research paper?

7. What are the four general elements of an APA reference?

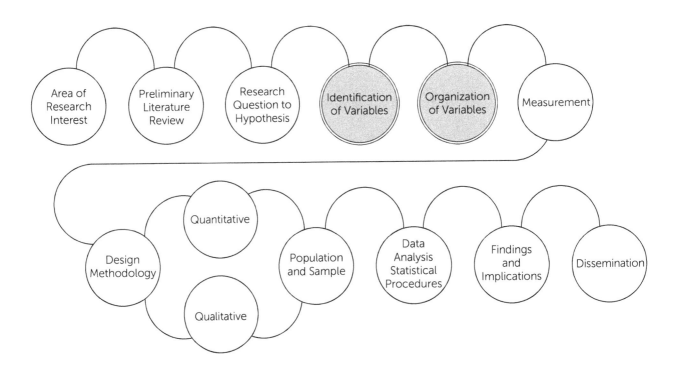

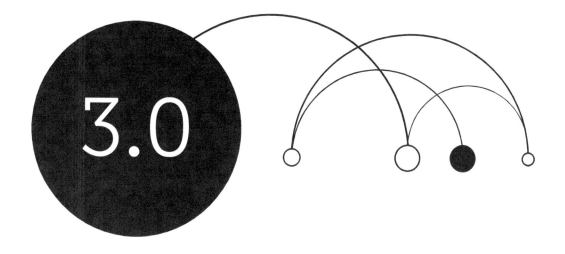

The Theoretical Framework of Research

To figure out how heavy a pig is, you find a good stout plank and balance it on the pole of a fence. Tie the pig to one end of the plank and then run around to the other side and put a rock on the opposite end. Keep trying different-sized rocks until you get one that balances the pig. Then all you have to do is guess the weight of the rock.

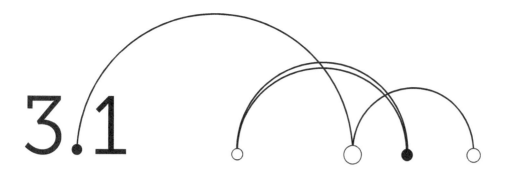

3.1

The aims of science: Putting research in context

Renowned researcher Fred Kerlinger provided several basic definitions that set the stage for structuring the process of research. He defined the broad endeavor of scientific research as the "systematic, controlled, empirical, and critical investigation of hypothetical propositions about the presumed relations among natural phenomena" (1973, p. 11). This fits with the aforementioned purpose of conducting research as contributing to the existing knowledge in a field. To do that, though, we must have a context—a framework if you will—on which to attach the new pieces of knowledge that our research uncovers.

This organizing structure is represented by the theories developed to provide overall explanations of the various pieces of amassed knowledge. Theory, according to Kerlinger, represents, "a set of interrelated constructs (concepts), definitions, and propositions that present a systematic view of phenomena specifying relations among variables, with the purpose of explaining and predicting the phenomena" (p. 9). If this is the case, we need to determine, through our review of the literature, where in this theoretical framework our new research will reside.

Eisenhart defined the theoretical framework as "a structure that guides research by relying on a formal theory … constructed by using an established, coherent explanation of certain phenomena and relationships" (1991, p. 205). Thus, your theoretical framework should lay out the specific theories and theorists upon which prior research is grounded and to which your new research will contribute. In doing so, your theoretical framework will encompass not only an explanation of your research question but also the details about your associated variables, with a specific emphasis on the rationale for choosing these in light of existing theoretical constructs.

From a practical standpoint, the theoretical framework provides a rationale and a level of justification to conduct a study. It gives a broader view of the purpose and significance of the research and provides context for how the new insights will contribute to the knowledge base of the field—which is, as you may recall, the fundamental purpose of research.

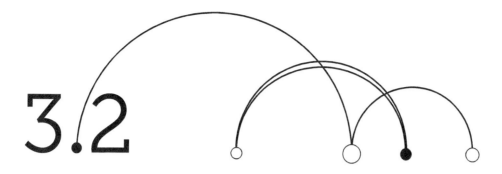

3.2

Variables and their types

A **variable**, as you might suspect, refers to something that varies or changes. In other words, it can take on different values, characteristics, or categories. The process of research involves observing the effects of certain variables on other variables or of describing (and predicting) the relationships between certain variables. These variables are, essentially, the objects of all this fuss and bother. Our research endeavors attempt to untangle the tangled web they weave in the larger picture (the theoretical framework) of understanding phenomena in the field of interest.

As such, you can imagine, they come in various types with differing characteristics. There are several ways to categorize these important players.

A fundamental way to define variables is to describe their function or role in the proposed relationships under examination. The *independent variable* (also called the predictor variable) is the variable that is observed or manipulated to determine its effect on a particular phenomenon. The *dependent variable* (also called the outcome variable or criterion variable) is that phenomenon. The independent variable is the characteristic or condition that is assumed to cause or influence a change in the dependent variable. The dependent variable is of primary interest to the researcher, who, through the process of research, seeks to understand and predict how this important phenomenon reacts to various changes and conditions.

It is important to note that any given variable can be either independent or a dependent. It is not an innate characteristic of the variable, but rather a function of its presumed role in the relationship under study.

While it is the underlying thinking that the independent variable is "assumed to cause" changes in the dependent variable as we set up a research study, caution is advised. You may remember from your statistics course that supporting a claim of causation is actually quite difficult (if not virtually impossible). Only in true experiments, under tightly regulated conditions with control over any and all other variables, can we even approach conclusions of causality when supported by the data. When we describe the relationship between the independent and dependent variable in this way, we are only referring to the mindset with which we set up our study (i.e., what we think is going on and, therefore, what we will be investigating further).

Concluding statements of cause and effect must be made only when our research design and statistical analysis warrant such bold declarations. Specifically, as outlined by Sekaran and Bougie (2013, p. 70), "to establish that a change in the independent variable *causes* a change in the dependent variable, *all four* of the following conditions should be met:

1. "The independent and the dependent variable should covary—a change in the dependent variable should be associated with a change in the independent variable.
2. "The independent variable (the presumed causal factor) should precede the dependent variable—there must be a time sequence in which the two occur: the cause must occur before the effect.
3. "No other factor should be a possible cause of the change in the dependent variable, so the researcher should control for the effects of other variables.
4. "A logical explanation—a theory—is needed, and it must explain why the independent variable affects the dependent variable."

While independent and dependent variables may be the sole point of a research investigation, it's usually not that simple or straightforward (of course not!). Because most phenomena are complex and multifaceted, usually more than one variable operates to influence any given dependent variable (at least outside the lab in the real world, and particularly with respect to human behavior). We may not be interested in these variables, but that doesn't mean they are not exerting an influence. Thus, we must take them into consideration.

Extraneous or *confounding variables* refer to other factors—whether they be individual, situational, or global—that are not the focus of interest for the study (i.e., are not being measured) but still may influence the dependent variable and, thus, the results of the study. Depending on the nature of your study—the variables, subjects, etc.—there may be a large number of extraneous variables, some of which may be difficult to control for.

Intervening variables (also called *mediating* or *intermediary variables*) refer to a construct or phenomenon that provides a causal link between variables. For example, in early experiments on learning and motivation, lab animals were deprived of water to determine the effect on the rate of lever-pressing to receive water. The hours of deprivation represented the independent variable, while the rate of lever-pressing represented the dependent variable. However, under further scrutiny, it can be argued that it is not the hours of deprivation that "caused" an increase in the rate of lever-pressing, but rather the intervening variable of thirst. In this case, hours of deprivation cause thirst, which in turn causes the increase in the rate of lever-pressing.

Moderating variables represent another type of variable that can occur in some research scenarios. The moderating variable changes—moderates—the relationship between the other variables, producing what we refer to as an interaction effect. While a mediating variable accounts for or explains the relationship, the moderating variable can affect the strength and/or direction of the relationship between the independent and dependent variable.

Referring back to our lever-pressing lab animals, thirst (the mediating variable) accounts for the relationship between hours of deprivation and rate of lever-pressing. However, imagine that female lab animals show a significantly larger increase in lever-pressing over their male counterparts. In this case, sex of the animal would represent a moderating variable because it affects the strength of the relationship (i.e., the rate). You can imagine that age could also play a similar role in these experiments. Thus, a number of moderating variables can be present in any study and must be considered with respect to the underlying cause-effect rationale.

The quantification of variables is another way in which we categorize different types of variables, and these distinctions have important consequences on the data analysis that occurs later in the research process. Data are usually numbers, so here we really are talking about the types of numbers that can form data.

Numbers come in four flavors, so to speak: **nominal**, **ordinal**, **interval**, and **ratio**. Let's take them one at a time. Figure 3.1 shows the types of data that will be covered over the next few pages and provides examples for each.

FIGURE 3.1 Measurement Scales

Type of Scale	Example
Nominal	Gender (male, female), Political party (Democrat/Republican/Independent)
Ordinal	Basketball standings, sib-lineposition (first born, second born...)
Interval	Temperature in degrees, intelligence test scores, SAT scores
Ratio	Age, height, weight, speed

Nominal data can also be called *dichotomous* (only two levels, like male and female) or categorical data, where there may be more than two categories (e.g., Republican, Democrat, Independent). It is a classification that has no inherent mathematical properties. For example, male and female are nominal (and dichotomous); Democrat and Republican are nominal (and dichotomous); Protestant, Jew, and Muslim are nominal (and categorical); absent-minded professors and schizophrenics are nominal and categorical (well, maybe that last example can be argued—it may be the same category).

Often, nominal data are assigned numbers for computer entry purposes. It is easier to enter a 1 for males and a 2 for females, or vice versa, than it is to type in "male" or "female." Even so, the numbers that designate categories for computer purposes have no mathematical properties other than to present the short-hand identification of the nominal category represented by the number. For example, if Republicans equal 1 and Democrats equal 2, and you add 1 and 2 together to get 3, what is a 3? A 3 has no mathematical meaning. You cannot say that a 3 represents Independents, because Independents are not necessarily the mathematical combination of Republicans and Democrats. One can certainly make Independents a 3 for computer recognition purposes, but that is just a categorical designation like 1 for Republicans, 2 for Democrats—it is not a number derived through any mathematical process.

As can be seen, no logical mathematical procedures can be applied to numbers that are used to represent nominal data. Designating nominal data by numbers is really just a matter of convenience. This is not to discount the importance of nominal data. Just think of all those studies that have been conducted looking for differences between the genders. How else would we know that men are from Mars and women from Venus?

Ordinal data (numbers) indicate order only, but may not indicate what measurement was used to determine the order or the magnitude of the differences within the order. For example, we might say Tom was first, Mary second, Dave third, and Harry last. This is ordered data, therefore ordinal, but we do not know what these individuals were first, second, etc., *on* (measurement) or how far ahead Tom was of Mary, Mary of Dave, or Dave of Harry. This is just a simple example to illustrate ordinal data.

In real life, we usually would not collect this kind of ordinal data—we would just measure each subject directly and get an actual number, but sometimes in research, all we want is order. For example, we may have several graduate students watching inpatients in a hospital setting and ranking (ordering) them on levels of aggression. We may not need an actual measure of aggression; that is, an actual aggression score. We may just want patients rank-ordered from most to least aggressive, at which point, we may assign the more-aggressive patients to intervention groups based on the rank-ordered levels of aggression.

For another example, we may have two assistants rating student essays from best to worst, rank-ordered only, and not need an actual "score." Or we may have blind raters (*blind* meaning they do not know what the study is about) categorizing individual statements from most psychotic to least psychotic. These would be situations where only ordinal data would be more likely to be gathered.

Interval data are true score data where you know the score a person made and you can tell the actual distance between individuals based on their respective scores, but the measure used to generate the score has no true zero. Most psychological measures fit into this category. For example, IQ test standard scores have no true zero (the IQ conversion tables do not go down to zero). With interval data, I can legitimately say that a person with an IQ of 100 has a score of 100. I can also say that a person with an IQ of 100 is 50 IQ points higher than a person with an IQ of 50. I cannot say that a person with an IQ of 100 is twice as smart as a person with an IQ of 50. One must have a true zero to generate such a ratio.

This brings us to the last type of data: **ratio**. As the name implies, it is data where ratios can be calculated. Most physical measures are ratio data, e.g., height, weight, speed, distance, volume, area, etc. Ratio data have all the properties of interval data and then some. For example, as with interval data, we can state the distance between measures on different individuals, so we could say that a child 4 feet tall is 2 feet taller than a child 2 feet tall. In addition, we can also say that a child 4 feet tall is *twice* as tall as a child 2 feet tall, thus stating the ratio. (A graduate student suggested to one of your authors that height data have no true zero—that at the moment of conception, there is height, even if it is just the height of a single cell. We placed that conundrum in the same category as how many angels can dance on the head of a pin and suggested the student might be more suited for philosophy than psychology.)

We hope it is obvious that ratio data are the most precise, with interval next, and then ordinal, and nominal not really qualifying for an assignment of precision since the categories are usually obvious (male or female) or stated by the research subject ("Yes, I am Republican."). Ratio data can be turned into interval. If you are familiar with IQ tests, you know that the raw scores calculated in response to the subject's actual answers (as on the Wechsler scales) are ratio (true zero), but are then converted into scaled scores, making them interval (no true zero).

Ratio and interval data can be turned into ordinal simply by dropping the numbers and adding a placement or rank designation. For example, if we know the actual height of five individuals (ratio data), we can make that ordinal by saying that Tom is tallest, John next, Billy next, then Mary, with Sally the shortest. We could even go to nominal data if we establish two categories, say, under 6 feet versus 6 feet and above. We could do the same if we had actual IQ scores (interval data) by giving order, low to high or high to low, and convert to nominal by creating two categories, perhaps an IQ of under 100 versus IQs of 100 and above.

We can work down the hierarchy from a higher level of precision to a lower level of precision, but we cannot go in the opposite direction unless, as with ratio and interval data, we have a table that converts the interval data back to ratio (one can do this with IQ tests by working backward in the conversion tables, although some precision may be lost since some interval scores on subtests of the Wechsler scales may represent a range of ratio scores, e.g., a scaled score of 12 on Block Design for 12-year-olds covers ratio scores from 48–51).

Generally speaking, one would not want to convert more powerful and precise data to less precise and powerful. We point out that it can be done simply for illustration purposes.

Other Characteristics of Data

Discrete and continuous numbers

Numbers also come in **discrete** and **continuous** form. Discrete is what you probably remember as whole numbers; that is, numbers without decimals. What you called *fractional* or *decimal* numbers are called *continuous* numbers by statisticians.

In statistics, all discrete numbers are assumed to have continuous properties. The continuous, or fractional, extension of a discrete number goes from half below the number to half above, so the number 100, a discrete number, has an assumed continuous range from 99.5 to 100.5. The discrete number 101 would share the **lower real limit** of 100.5 with the **upper real limit** of the discrete number 100. In this sense, each discrete number bumps into its neighbor at half below, half above, as shown in Figure 3.2.

FIGURE 3.2 Sample of Discrete Numbers

You might suggest that having shared lower and upper real limits creates error. You would be right. The upper limit for 15, in the example above, is really $15.4\overline{9}$. By the time you get out to infinity, which you will never do, theoretically, the difference between $15.4\overline{9}$ and 15.5 would be so small that the error of the shared limit of 15.5 between 15 and 16 would be negligible.

Even if a number occurs only in discrete format in the real world, it can be continuous in the statistical world—and that would be acceptable. As an example, when a psychologist administers an intelligence test, the IQ scores are always expressed as discrete numbers. It would make no sense to have IQ scores such as 115.68 or 123.24. That would certainly imply a precision to the measurement of IQ that is not there. However, when using IQ scores in research, it is entirely possible for the combination of IQ scores, such as the average of several scores, to be continuous. If this happens, it is fine to leave the number as continuous—it does not have to be rounded back into a discrete format.

This is about all we need to say about types of numbers at this point, and it is about all you need to know from this point forward regarding types of numbers. Recognizing the type of number is important because, if you have had a course in parametric statistics, you know that formulas can change depending on the type of variable you are dealing with, whether nominal, ordinal, interval, or ratio.

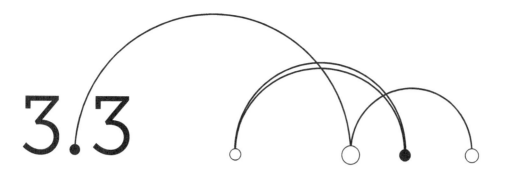

Operationalizing variables

Variables represent the functional components of research questions. They are the items on the table for study. It is the determination of their relationship (or lack thereof) that will comprise the rest of the research endeavor—that is, as soon as we get them operationalized. To move forward in the process of research, we have to take our variable—a concept—and turn it into something that we can measure. As Kerlinger put it, the operational definition "assigns a meaning to a construct or a variable by specifying the activities or 'operations' necessary to measure it" (1973, p. 31).

The measurement of some variables is relatively concrete and obvious. Height is measured in feet and inches (or meters and centimeters). Weight is measured in pounds and ounces (or kilograms and grams). Temperature is measured in degrees. Time is measured in years, days, hours, minutes, and seconds, etc.

Other variables, especially those related to human behavior, can be more intangible or difficult to define specifically. Personality, intelligence, motivation, anxiety, etc., are all familiar concepts, but pose a bit of a challenge to clearly define in ways that are measureable and comparable among individuals or groups. Each of these concepts has existing formalized measurement instruments—the Myers-Briggs Type Indicator (MBTI), IQ score on the Wechsler Adult Intelligence Scale, Murray's Manifest Needs test, the Generalized Anxiety Disorder Assessment (GAD-7)—and each of these has gone through rigorous validly and reliability testing. However, these tests represent only one of a variety of measures for their associated concepts. Additionally, each instrument typically has both proponents and opponents (i.e., those who agree that the test is a comprehensive measure of the underlying construct and those who do not agree).

Still other variables fall somewhere in the middle. Productivity, for example, can be measured quite concretely if we are referring to the number of widgets produced within a period of time. But what about measuring productivity of a less-tangible product, such as creativity or management skill? What about effectiveness? We know that being effective has different meanings depending upon the context. An effective teacher performs differently from an effective electrician and still differently from an effective pilot.

We can come up with various ways to measure this somewhat elusive "effectiveness" for each of these contexts, but clearly the behaviors (i.e., the measures) will be different—that is, if they are a valid measure for that particular context (see the discussion on validity in Section 4.5 of the next chapter). If we have chosen well, we wouldn't typically expect a teacher to perform as effectively as an experienced pilot on the measure we chose for pilot effectiveness.

Operationalization of variables (other than those concrete dimensions such as time and temperature) inevitably requires determining the concept's behavioral component—because behavior is what we can measure. However, a single behavior rarely encompasses the full picture of a broad concept. Additionally, the same observable behavior can have different underlying causes. For example, in the case of motivation with respect to a job, operationalizing "motivation" involves thinking about what behaviors a motivated person displays (such as being on time for work, seeking out new projects, etc.) and, in contrast, what behaviors an unmotivated person displays (being frequently late or reporting boredom).

You can already begin to see the challenge that arises as we start to apply measureable behavior to a more-abstract concept. Logically, some people are chronically late because they are not motivated by their work. However, not everyone who is chronically late is unmotivated. There may be other reasons for the lateness that have nothing to do with motivation. (In this case, these reasons would constitute extraneous or confounding variables).

Operationalizing variables is always an inexact science. Something is always lost in translation because single behaviors rarely have one consistent, traceable cause for everyone. Thus, there will always be a margin of error in our measurement.

In most cases, it is a good idea to use standardized measures that are already developed for your variables, if they exist. These measures have typically undergone testing for validity and reliability and have often been revised and improved over time. If you choose to develop your own measures, you will have to conduct pilot studies to ensure that they are adequate and appropriate means of assessing the chosen construct. This is further illustrated in Section 4.5 of the next chapter.

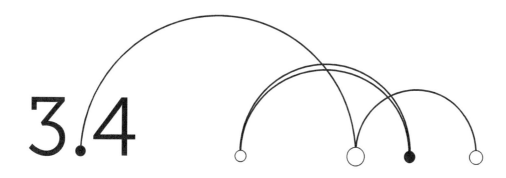

3.4

The hypothesis as a proposed answer to the research question: What you think is going on with those variables

The hypothesis represents the researcher's expectations about the relationship between the variables of the problem, or, as Kerlinger puts it, "a conjectural statement, a tentative proposition, about the relation between two or more phenomena or variables" (1973, p. 12). As we have seen, a research investigation begins with a question, but questions cannot be tested directly. It is the proposed relationship between the operationalized variables that is tested.

The hypothesis represents an educated guess about what will happen in an experiment in the form of a testable statement derived from the research question. It is considered tentative because it is the starting point for the investigation and will be tested empirically. As such, the hypothesis is derived directly from the research question and should predict a relationship between the identified variables in a specific and testable way. The hypothesis, thus, brings a clear focus and direction to the next steps of the research process: gathering relevant evidence (data) and analyzing that evidence to draw conclusions about the validity of your hypothesis.

By traditional conventions of scientific research, the logic of the hypothesis is divided into two opposing segments: the null hypothesis and the alternative hypothesis. In statistical hypothesis testing, these two possibilities are compared.

The null hypothesis (H_0) states that there is *no* relationship between the two variables being studied (one variable does not affect the other). It is a negation of what the experimenter actually expects or predicts. A null hypothesis is used because it enables researchers to compare their findings with chance expectations through statistical tests. It states that the results are due to chance and are not significant in terms of supporting the idea being investigated.

The alternative hypothesis (H_1 or H_A) states that there *is* a relationship between the two variables being studied (one variable has an effect on the other). It states that the results are not due to chance and that they are significant in terms of supporting the theory being investigated.

Research hypotheses may be either nondirectional or directional. A directional hypothesis specifies the nature of the relationship or difference that is predicted, so a specific group or condition will be higher or lower, or have more or less of something, *or* it predicts the specific direction that a correlation will take. A nondirectional hypothesis, on the other hand, states that a relationship or difference exists, without specifying the nature of the expected finding.

An example of a directional hypothesis would be: Deaf children whose parents are deaf will acquire language faster than deaf children whose parents are not deaf. This specifies the direction of the condition: One group is faster than the other.

An example of a nondirectional hypothesis would be: Male and female supervisors will differ significantly in their use of threats as a compliance gaining strategy. This indicates only that there is a difference between the groups, but doesn't indicate the direction (i.e., which group uses more threats in its strategy).

Suppose our general research question is, "Do students work more efficiently on Monday morning than on Friday afternoon?" From this starting point, we need to operationalize the variables to be able to go out and gather data to test our assumption. We could decide to study this by giving the same group of students a lesson on a Monday morning and on a Friday afternoon, and then measuring their immediate recall of the material covered in each session. In this case, our hypothesis could be: "Students will recall significantly more information on a Monday morning than on a Friday afternoon."

Note that stated this way, we have a directional hypothesis. We can see how this hypothesis flows clearly and directly from the research question with only the addition of a few specific details to operationalize the way in which we would go about investigating this question.

For the research question "Does eating breakfast affect student performance?", a hypothesis could be, "Students who eat breakfast will perform better on a math exam than students who do not eat breakfast."

For the research question, "Does using a cellphone while driving affect driver performance?", a hypothesis could be, "Motorists who talk on the phone while driving will be more likely to make errors on a driving course than those who do not talk on the phone."

Note that each of these examples represents one possible hypothesis (and one way of operationalizing the variables) that could be derived from the associated research questions.

Numerous other, equally valid, possibilities exist that could test the same general research questions. In addition, notice that each of the testable hypothesis statements also has the potential to incorporate some extraneous variables that could influence the results. Can you identify some for each? Remember that these must be addressed as the study progresses to have confidence in the analysis of our results.

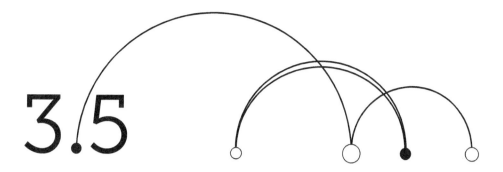

The rationale of no difference: The null hypothesis

As stated previously, by convention, we create a null hypothesis that is essentially the opposite conclusion of what we expect to happen as we observe the interaction of our variables. Again by convention, it is the null hypothesis that is tested. This may seem a bit counter-intuitive and awkward, but it is based on the philosopher Karl Popper's Principle of Falsifiability, which states that we can't conclusively confirm a hypothesis, but we can conclusively negate one. Basically, with respect to research, the idea is that no theory can be completely proven correct, but if not falsified, it can come to be accepted as truth. Therefore, following this logic, it is not possible to test a hypothesis directly. Instead, you must turn the hypothesis into a null hypothesis, as described previously. The null hypothesis can usually be created from the hypothesis by adding the words "no" or "not" to the statement. The null hypothesis is, therefore, the opposite of the experimental hypothesis in that it states that there will be no change in the variable under study.

A familiar analogy for this is the criminal trial. Imagine you are the prosecutor. You believe, and want to show evidence, that the defendant is guilty. That is your alternative hypothesis. But you must assume that the defendant is innocent until proven guilty, so the null hypothesis is that the defendant is not guilty.

Let's look at some examples for creating the null hypothesis. From our previous example where our research question was "Do students work better on Monday morning than they do on a Friday afternoon?", we developed an alternative hypothesis: "Students will recall significantly more information on a Monday morning than on a Friday afternoon." In this case, our null hypothesis would be, "There will be no significant difference in the amount recalled on a Monday morning compared to a Friday afternoon." The null hypothesis assumes that any difference in the sample will be due to chance or confounding factors.

In our second example, our research question was, "Does eating breakfast affect student performance?" and our alternative hypothesis was, "Students who eat breakfast will perform better on a math exam than students who do not eat breakfast." Therefore, the null hypothesis would be, "There will be no significant difference in the performance of students who eat breakfast as compared to students who do not eat breakfast."

Finally, in our third example, the research question was "Does using a cellphone while driving affect driver performance?" Our alternative hypothesis was "Motorists who talk on the phone while driving will be more likely to make errors on a driving course than those who do not talk on the phone." Our corresponding null hypothesis would be, "There will be no significant difference in the number of errors on a driving course for individuals talking on a phone as compared to individuals not talking on a phone."

All statistical testing is done on the null hypothesis ... never the hypothesis. The result of a statistical test will enable you to either "reject the null hypothesis," or "fail to reject the null hypothesis." We don't use "accept the null hypothesis." Again, this traditional terminology is the convention used in scientific research based on the premise of falsifiability.

The null hypothesis is "supported" if the results are statistically nonsignificant. The null hypothesis is never "proven" (at least not by a single study) because of the impossibility of proving a negative. The null hypothesis is "rejected" in favor of the experimental hypothesis if the results are statistically significant.

In the next steps of the research process, you will proceed to testing the hypothesis. You will consider how to collect the data on the variables you have defined in your research question and test your hypothesis about their relationship. After collecting the data, you will analyze it to determine whether (or not) the evidence supports your hypothesis.

Where to Next?

In this chapter, we have provided you with the theoretical framework for research. Once you have your research question locked in, then you are ready to start. Well—almost. Before you can start, you have to identify where you are going to get your data, and which variables are important. Guess what? That is what we will discuss in the next chapter: variables, data sources and samples.

Key Terms

- Alternative hypothesis
- Categorical variable
- Dependent variable
- Extraneous variable
- Hypothesis
- Independent variable
- Mediating variable
- Moderating variable
- Nominal
- Null hypothesis
- Operationalization
- Ordinal
- Ratio
- Theoretical framework
- Variable

Questions

1. What is the significance of the theoretical framework?

2. Why is it necessary to operationalize variables?

3. Explain the relationship between the dependent and independent variables.

4. Describe the types of data and their significance.

5. What is the purpose of the null hypothesis? How is it different from the alternative hypothesis?

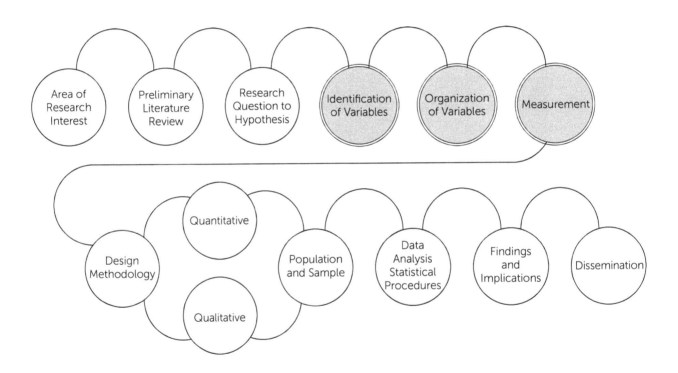

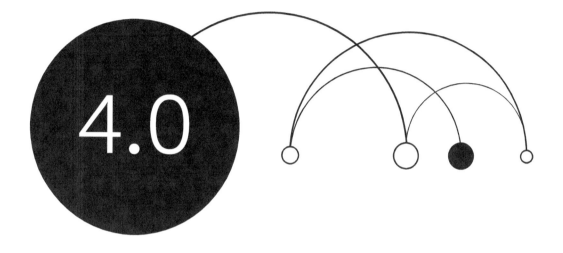

Identifying Variables, Data Sources, and Samples

A physicist, chemist, and researcher are sitting in the dean's office. Suddenly, a fire breaks out in the trash can. The physicist immediately starts working on how much energy has to be removed to stop the fire. The chemist starts working on a combination of reagents to put out the fire. The researcher, meanwhile, starts a fire in all the other wastebaskets. "What are you doing?" the others shout. "Well, to solve the problem," says the researcher, "you obviously need a large sample size."

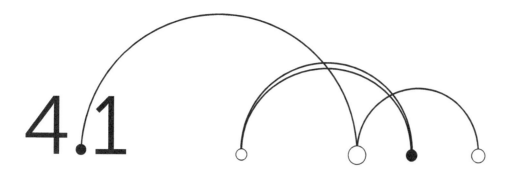

4.1

From research question to variables

Let us suggest that you are interested in the development of the skills needed to pilot an aircraft. That statement of interest automatically should send you to a body of literature that deals with piloting and training/skill development. As you peruse that body of literature to see what is known, what is not known, or what is suspected but not verified relative to piloting skills, you run across some early research suggesting some differences between males and females in the rate of skill development. This really triggers your interest, so you narrow down your literature reviewing process to piloting skill development and gender differences in skill acquisition.

As you review this new, narrower body of literature, again, looking for what is known, what is not known, and what is suspected but not verified, you run across an article suggesting that the differences in skill development between males and females may be related to the right parietal lobe of the brain (assumed to be less-developed in females). You find that *very* interesting! You recall some evolutionary speculation that spatial abilities (right parietal functioning) are greater in men because of their need in our dim past to hunt and maneuver in space in concert with other hunters, while females spent their time gathering nuts and berries and interacting with offspring. You delve more deeply into the literature dealing with parietal differences between males and females related to spatial skills.

What you have just done is taken an area of interest, read some material about that area, and, in the process, narrowed down that interest to a specific focus. It is the process of narrowing down from a broad field of knowledge to a more-specific field of knowledge that eventually leads to a research question. Your research question at this point is: Is there a difference in right parietal functioning between males and females that might affect learning to fly an airplane?

Your research question exists at the level of theory-model-construct-concept. That is, you ask questions related to the neuroscience theory of the function of the parietal lobes, and you deal with concepts such as differences and functioning. At this point, there is very little you can do in the way of an actual research study. You have some important questions to answer, such as:

- How can you assess parietal lobe functioning?
- How can you measure differences in parietal lobe function between the genders?
- How will you relate any differences, if found, to patterns of learning to fly a plane relative to the two genders?

To get to the point where you can actually subject your curiosity-driven research question to some analysis, you need to identify measures or create a means by which you can quantify parietal lobe functioning, assess gender differences, etc. Only then can you address your research question.

How will you measure right parietal functioning? To summarize from Section 3.2, a **variable** is a characteristic of some event or phenomenon that takes on different values. In short, a variable varies—changes—across whatever is being measured. Your living room will not measure the same as ours. Your car may have a different weight from ours. We can measure living rooms in square feet and cars in pounds, and there will be variability across the measures. Therefore, living room square footage or car weight could be a variable.

We can think of all kinds of variables: speed, grade point average, intelligence quotient, material density, dimensional vibrations in space measured by an interferometer. Pretty simple. What you have to do is turn parietal lobe function into a measurable variable—*not* so simple!

Other variables in your study may be a bit different from the parietal lobe variable or some of the other variables mentioned above. One variable in your study, of course, will be gender, and it is different. It is a **dichotomous variable**, meaning it has only two categories or levels: male and female. Dichotomous variables contrast with other categorical variables called **polytomies**— variables with more than two levels or categories, such as Republican, Democrat, Independent. The gender variable will be easy to identify; that is the least of your worries.

Next, you must wrestle with how to measure right parietal lobe functioning. You start looking around and find that the Halstead-Reitan Neuropsychological Battery may have some utility in measuring right hemisphere functioning (you will have to look into that further to see if it will do what you want). There is also the Rey-Osterrieth Complex Figures Task (a possibility). Subtests on some adult intelligence tests may give you what you need. Perhaps you could take a serial reaction time task and adapt it to your purposes.

There is always brain imaging if you have access to the necessary equipment (you give men and women parietal tasks and watch to see who lights it up most). If you are really fortunate, you may have access to a flight simulator and can design some tasks that will get at what you want. These are some possibilities for you to consider. Which of these possibilities—or others—might provide you with the measure you need to differentiate, accurately and reliably, right parietal function between the genders?

You need to jump into the literature, this time looking at parietal lobe functioning rather than pilot training. (Many students tend to see the "literature" as something that just has to be presented as part of a final paper. We hope, at this point, you are beginning to see the literature as your best friend, helping you clarify research questions, get a handle on potential measures, etc.). If there is a good measure of right parietal functioning, it will be out there in the literature. There is no need for you to reinvent the wheel.

Once you have settled on a reliable and valid measure, you will be on your way. First, restate your general research question from earlier as a **null hypothesis** with the measure included. That restatement might go as follows: "There *will be no* difference between male and female pilots-in-training on the Rey-Osterrieth Complex Figures Test." Acceptably, you can do it as an **alternative hypothesis**: "There *will* be a difference between male and female pilots-in-training on the Rey-Osterrieth Complex Figures Test."

Next, find subjects—randomly selected, if possible—and form two groups: one of randomly selected males in flight training, the other the same but composed of females.

"But, wait," you say. "I also stated that I wanted to look at the difference in parietal functioning, if any, and its impact on learning to fly." Here is where we come back to the characteristics of a good research question: parsimonious, precise, objective, operationalized, and rational (Section 1.7). If you add on this extra bit, are you going to overly complicate your study? It seems best to first determine whether there is a difference that is detectable before jumping ahead to see if it affects learning to fly.

It might be best for the learning-to-fly part to wait until you have determined there is a difference and that you are indeed measuring that difference accurately and reliably. Remember, research goes one small step at a time. First step: Is there a difference? Next step: If you find a difference, does it affect learning to fly?

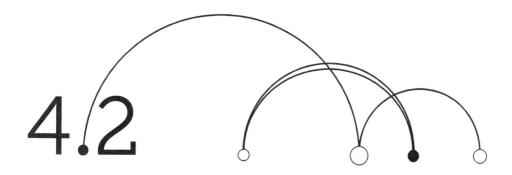

4.2

From variables to data sources

Continuing with this study, we know you need male and female pilots-in-training, hopefully randomly selected. Notice how your research question has defined your data source: who you will get your data from. Two of your authors have been looking at gender differences in crash codes using the National Transportation Safety Board (NTSB) database. That is an **archival data** set—already existing and available for public use. There are many such databases around. However, it does not look like there would be such a database for your particular study, so you are stuck with finding subjects.

We might suggest that pilots-in-training represent a relatively small number of people. Where might you find them? Private training schools might be one source. Aeronautically based academic programs that include such training as part of the curriculum might be another. Certainly, the military does a lot of training. These all represent possible sources for your data. You must address the feasibility of each:

- Is it likely the military is going to let you come onto a military base and measure their trainees?
- Is it likely that private instructors are going to give up airtime with students so the students can participate in your study?

These are all relevant questions you must find answer to before you know where you will get your data.

"Wait a minute," you say. "How about if I just collect data from members of pilot associations?" There are a lot out there and it might be the quickest and easiest way to go. We might ask whether there are any drawbacks to doing that? How many can you think of? In the context of this study, the primary drawback would be that these are experienced pilots. Would experience override any differences you might see between the genders in novice pilots-in-training? That is a good possibility.

You decide to head to academia for the greatest concentration of pilots-in-training. Now you have to look at the curriculum and decide where in the training process you need to collect your data.

As you can see, narrowing down your source is not as easy and straightforward as you might have thought. Additionally, how you do that may raise questions about the veracity of your data. If you use freshman trainees, are you too early (they have not really started actual flying yet)? If you use senior trainees, are you too late (they are well advanced and experience is beginning to negate differences)? Many a research study has been undone because the data source was questionable in some way.

We hope you can see, at this point, how the research question necessarily defines your data source. Without that research question, you will have no idea where you need to go.

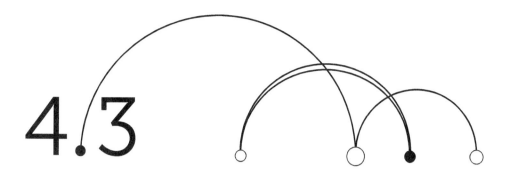

4.3

From variables and data sources to samples and populations

You have decided to use sophomore pilot trainees in your study. This seems to be the ideal level, where there is enough training to classify the subjects as "pilots" but still ensure they are beginners and learning basic skills. You will want to find a number of such academic training programs and, if there are enough sophomores across the institutions, you will want a random selection of subjects.

Random sampling distributes potentially important and confounding variables evenly and normally across your sample, thus mimicking the distribution of those variables in the population, making your sample **representative** of the population. **Confounding variables** are ones that you did not control for in your study and can contribute to an explanation of your findings. For example, suppose the average age of your male pilots-in-training is 32 and the average age of your female-pilots-training is 24. Would age contribute to a neurological maturity that might explain any differences you find? We do not know, but it seems a likely explanation and, unless controlled, would be a confounding variable.

Your randomly selected group of sophomore pilots-in-training is your **sample** (called a **probability sample**). The **population** of this sample is all sophomore pilots-in-training. Note we did not say "all pilots-in-training." You cannot extend your findings, should they be worthy of extending (meaning significant), back to all pilots-in-training. You can only extend your findings back to the population of sophomore pilots-in-training and only then if you did indeed sample randomly from a number of academic institutions. If all of your subjects came from the Yale Divinity School of Pilot Training (a wing and a prayer), then your findings apply only to the other students in that particular program. Note that populations can be quite circumscribed. *Population* does not necessarily mean everybody in the world.

A quick note about random sampling. When we say that it creates a sample that mimics the population, that is an assumption, not a fact. There is something called ***random error***, where random sampling does not create a sample mimicking the population. Generally, however, if the samples are large enough, and sampling is conducted as it should be (every subject has the same chance of being selected each time a name is drawn, meaning that each name drawn goes back in the hat rather than being discarded), the assumptions of random sampling can be quite robust. You must realize, of course, that there is always some degree of uncertainty. However, as researchers are fond of saying, "Uncertainty is a certainty of life."

Notice that we are getting into some strict rules about how research is conducted. It is these rules, which have been developed over the years, that have raised research above the level of speculation, biased observation, anecdotal evidence, conclusions based on experience, and opinion, to the level of scientific endeavor.

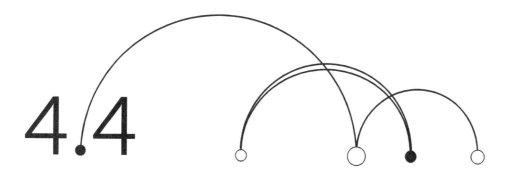

4.4

Populations and samples: The process of extrapolation

From Chapter 3, you know that the goal of most research—we might say *all* research—is to be able to apply findings beyond your sample, called ***extrapolation***. If you perform a study using the kids in your neighborhood, any findings will apply only to the kids in your neighborhood. To have a wider applicability, your sample should be a random selection (called a ***probability sample***) from a definable population.

If the population is definable, but your sample is not random, then you have some additional problems with extrapolation back to the population. These nonrandom samples are called ***nonprobability samples***. While lacking some of the research virtues of random samples, thus putting constraints on your procedures and findings, nonprobability samples do have a place in research. The NTSB database mentioned earlier is a nonprobability sample. Fortunately, it is also the population.

One of your authors was involved in research on childhood depression back when it was thought that children evidenced depression, behaviorally, the same way as adults. Working in a state psychiatric facility, we evaluated, measured, interviewed, and classified every child coming into the hospital. That was not a random sample by any means. However, over the course of about five years, working with researchers at other institutions spread throughout the eastern seaboard, we were able to generate a large database from which we could begin to select random samples. While we certainly reported findings at conferences and in the literature before random sampling, our findings based on random sampling had more authority to them and required fewer qualifications in presenting those findings.

The ability to apply your findings beyond your sample is only as good as the choices you have made in conducting the study. You do have choices at each step: from the adequacy of your review of the literature to develop your research question and measures, to the broad research design you select (univariate, multivariate, parametric, or nonparametric), to the accuracy of your data gathering, to the reliability and validity of the measurements chosen, to the adequacy of the statistics or qualitative methods used to analyze the data.

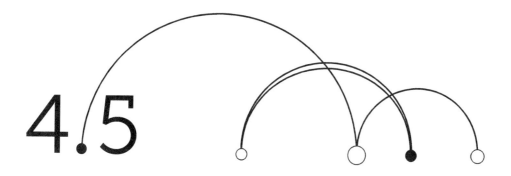

4.5

Essential characteristics of variables: Reliability and validity

As we continue with your pilots-in-training study to illustrate reliability and validity, **_reliability_** simply refers to the consistency of measurement. Remember that measurement is the assignment of quantity to some event, object, or phenomenon following rules. Here, we come back to rules again.

Reliability

Reliability is a rule of measurement. You have to measure accurately and consistently to have useful data. This a basic rule of science, whether you are measuring something or observing something. If you observe someone who one day acts politely, nicely and considerately, and the next day acts loudly, abusively, and self-centered, there is an inconsistency in your observations. If you measure somebody's height today as 5 feet, 10 inches, but tomorrow that same person measures 6 feet, 4 inches, then you have an inconsistency in your measurement.

It is perhaps easier to explain the inconsistency in the mood of the individual than in the height variation. In the case of the person's height, your data are obviously lacking in accuracy and are inconsistent, therefore lacking in reliability. If two people were observing the same person at the same time, and one saw that person as polite, nice, and considerate, and the other saw the person as loud, abusive, and self-centered, then, again, you have inconsistency in the observation and no reliability across the observations.

In your pilot training study, if you are using a well-known and standardized measurement instrument, such as the Rey-Osterrieth figure, you are probably okay on reliability. If you have made your own measurement, perhaps judging lengths of moving lines on a screen with just the left eye, you will need to establish reliability (consistency of measurement) first.

As an example, let us suppose that one of your authors wishes to construct an intelligence test. We will call it the Politano Intelligence Test Scale—PITS for short. There are initially a number of different considerations in developing this test. The construct of intelligence is very broad and contains many facets. It has been proposed that there is an artistic intelligence, a kinesthetic intelligence, a social intelligence, an academic intelligence, etc. (Gardner, 1993). Which aspect of the construct of intelligence am I interested in measuring? How will I make my test unique and more marketable than other tests that purport to measure the same thing? After all, I do want to make money from this enterprise and, if my test has no advantages over others, then it will have little market appeal.

Let us say I finally decide to go after academic intelligence. I am interested in determining who might do well in college, graduate school, etc. I develop the PITS with a unique feature: It only takes 15 minutes to give. That is my marketing catch.

After much work (at least an hour or two), I develop the test and decide to give it to several classes of my students. I spend most of a day on the first of October giving it to students and getting scores for each individual. I then readminister the test in late December to the same students. I want to see if their scores remain about the same over that period of time—that is, does the PITS have reliability (consistency) in measurement?

On my first administration of the PITS, Suzy Q receives an IQ score of 137. On my second administration, a couple months later, Suzy Q has a score of 89.

Two possibilities could explain this change in Suzy's scores. First, Suzy Q's brain could have started to rot. Second, my test may lack consistency in measurement. If the change in scores for Suzy Q is due to a lack of consistency in measurement, that would mean that my test does not have reliability.

To repeat, reliability is the consistency with which something is measured. Reliability is absolutely necessary for any measurement that is worth anything at all. Without reliability, you have data that fluctuate. If the data fluctuate simply because of an inconsistency in the measurement itself, apart from other factors, then the measurements are not reliable and the absence of reliability essentially means that you have nothing in the way of meaningful data. In other words, you have garbage.

Imagine how chaotic the world would be without reliability. Mile markers on the interstate might vary greatly in terms of the distance between them. You might live in Chicago but be closer in miles to Los Angeles than Detroit. Your baby might grow to 6 feet in just a few months! One day you may be dangerously overweight and the next day on the brink of starvation, given your weight. Perhaps these are exaggerations, but you get the drift. Without consistency in measurement, life would take on a considerably greater amount of uncertainty.

Now I must go back to work on my test and see if I can increase the consistency of measurement, i.e., increase reliability. I may add some items, take out others, clear up ambiguities on some, etc.

After a time, I get the PITS to the point where it seems as if it might measure more consistently because of all the changes I made to the items. I also increased administration time to 20 minutes with new items, but that is still very short for this type of test.

I have a new statistics class by now and give them the revised PITS at two different times. What good fortune: It seems to be measuring consistently, and there seems to be reliability of scores across time.

On the first administration of the revised PITS, Sally Jo's IQ score is 121; on the second, 122. All the other students' scores are very close on their first and second administration scores. Those close and consistent scores now suggest that my PITS has been revamped enough that there are some initial indications of reliability. Oh, what joy—I shall be rich, rich, rich!

Validity

Validity is whether your test measures what you think it is measuring. Now I want to make sure my test has validity and is as good as other tests already out there that measure academic aspects of intelligence, but take much longer. I again administer the new and improved PITS to my students, along with another well-recognized measure of academic intelligence that takes more than an hour to administer. I want to compare the students' scores on the PITS to their scores on this other test. I note that, on my revised PITS, Mary Sue has an IQ of 126, but on this other IQ test, her score is 113. That is a 13-point difference—a very large spread in the realm of IQ scores.

I examine the scores on the two tests for my other students and note that all of their PITS scores are considerably off from this other well-recognized academic intelligence test—in some cases, higher; in some cases, lower. This could suggest that my PITS may be measuring something a bit different from this other test, and it raises the issue of validity.

For the PITS, I wanted to measure academic intelligence. Is that what I am measuring? That, at its simplest, is the definition of *validity*: Are we measuring what we intended to measure?

I constructed the PITS to measure academic intelligence, but it does not seem to be measuring that aspect of intelligence in comparison to this other well-accepted measure of academic intelligence. Maybe I am measuring something other than academic intelligence or including material in the PITS that is related to other aspects of intelligence. Perhaps I have defined "academic intelligence" a bit differently. Maybe my test is too short to measure the construct of academic intelligence adequately.

The research on this other test, which has been around a long time and is well accepted in the professional community, clearly indicates that it measures academic intelligence—that is, it is a valid test for academic intelligence. It will be hard for me to convince people that this long-used test has been wrong all these years and that my shorter test, without the years of validating research, is the correct measure of academic intelligence. Therefore, I must conclude that my test may have questionable validity—specifically, it may have questionable **construct validity**.

Perhaps I am measuring some aspects of social intelligence or kinesthetic intelligence, when I really wanted a pure measure of academic intelligence. When you think about it, constructs such as intelligence are often complex and may contain many facets. You have to be sure which facet you are measuring—which piece of the construct is being addressed.

Here is an important point about reliability and validity: *If you do not have reliability, you cannot have validity!* That should be common sense. If you cannot measure something accurately, then you do not have data worth using and, therefore, it cannot have validity.

Let's suppose you measure a sample of participants' heights over a three-day period and get different results, such as:

- On day one, Suzy Q's height is 5 feet, 6 inches.
- On day two, her height is 4 feet, 10 inches.
- On day three, her height is 6 feet, 4 inches.

Several things may have happened. First, Suzy Q could be going through some dramatic growth spurt. Second, you may have lost your glasses and cannot see beyond your nose. Most likely, you have just messed up your measurement and have no reliability regarding those three days of measures relative to Suzy Q. That means that Suzy Q's height data are garbage! Without that reliability, you do not have valid data height about Suzy Q.

I must now examine my items specifically to see if they only address academic intelligence. I will make some more modifications, weed out some items that do not appear to be tapping into academic intelligence, and add some others. As an extra measure, I am going to give the PITS, after this latest round of modifications, to some colleagues who are skilled in intelligence testing. I tell them about the PITS and ask them to look at it with their expert eyes and give me a read on whether it looks to be a good measure of academic intelligence.

They look at it and say, "Well, it may be measuring academic intelligence—kind of hard to tell—but it looks that way." That is not exactly a resounding affirmation. Their ambivalence suggests two possibilities. The first is that they are not the experts that I supposed them to be in the first place (I like this possibility). The second is that my test lacks **content validity** (I do not like this possibility). Content validity is when a test appears to be measuring a defined construct (in this case, academic intelligence) to experts in that particular area. Since they are experts, they are giving me their knowledgeable opinions on the content of the test relative to the assessment of academic intelligence.

I make a few more modifications in about 30 minutes' time. Rather than spend more time on making more modifications, I decide to administer the slightly modified PITS again to another class, along with my criterion measure (the well-accepted academic intelligence test). Doing this without more extensive modifications in light of my colleagues' appraisals may seem hasty, but I am in a hurry to become rich and famous.

What I am hoping for this time around is that my scores on the PITS maintain a close and consistent pattern with my criterion test—the well-recognized and accepted measure of academic-related intelligence. If so, then my scores would converge on the scores of the other test and I would have **convergent validity** on the construct of academic-related intelligence. Convergent validity is an important aspect of demonstrating construct validity.

Suppose, hypothetically, that my test scores do not converge with the scores on my criterion IQ test but, instead, overlap with a well-accepted test that measures occupational aptitude—that is, whether a person is inclined toward science fields versus more artistic endeavors, etc. Oops! The PITS was supposed to be an academic-related IQ test, not an occupational aptitude test. As an IQ test, the scores should diverge somewhat from scores on an occupational aptitude measure. When scores on a test do not overlap with scores on tests that measure other, and dissimilar, constructs, that is usually good and represents another aspect of construct validity called **discriminant validity**. You want your test to converge on tests that measure the same or a similar construct—convergent validity, and not overlap with tests that measure divergent, or different or dissimilar, constructs—discriminant validity.

Now I must work the test around again, to increase the overlap with my criterion for academic intelligence measure and reduce, if not eliminate, its overlap with tests that measure other more remotely related abilities or aptitudes—measures that are dissimilar to some degree. Then I will be getting close to demonstrating that my test addresses the specific construct that I wanted to measure—that is, academic-related intelligence. If I get the test to this point, the PITS will begin to demonstrate convergent and discriminant validity, both of which will suggest construct validity. If I give it to my colleagues again and they like what I have done, it will begin to demonstrate content validity.

I take one more stab at reconstructing the test. It now takes 35 minutes to administer, which may reduce some of its marketability, but a test lacking appropriate psychometric properties such as reliability in particular and validity in its various forms would not have much marketability anyway.

I decide to take my reworked test and administer it to a small sample of students, just to see how it does. I will do a larger study, comparing it with the criterion test after this small sample. I give it to 10 students, telling them that it is a measure of academic intelligence. While administering the test, I notice that most, if not all, of the students are snickering at some of my items. I ask them what is so funny. They say, "You have got to be kidding. This is an intelligence test? It looks more like some of those hokey tests you find online that measure your astrological signs or something like that." My test obviously does not look like an IQ test to them.

Perhaps my other students who have been my guinea pigs while all this has been going on have been saying the same thing: "This is an intelligence test? Are you sure? These questions do not seem to have much to do with intelligence to us." If the test does not look like an intelligence test to the people taking the test, then it lacks **face validity**. If I ask somebody to take a test that is supposed to measure something specific, and it does not look like it measures that specific area to the test-taker, then the test-taker might not take the test seriously because it looks to be such a mismeasure of what it is supposed to be measuring. They may just blow it off as a waste of their valuable time.

What to do now? Continue working on it? I think not. After all, the test was the PITS to start with.

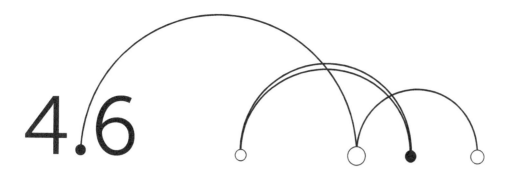

Almost beyond the shadow of a doubt: Probability and chance

Probability is a simple concept. It is betting, odds making, gambling, taking a chance, risk assessment, carrying an umbrella on a potentially rainy day, etc. If you watch horseracing or bet on horses (or on anything), you are engaged in probability: What is the probability you will win? If you watch the weather, you are engaged in probability: How necessary is it for me to carry my umbrella today to work or school—that is, what is the probability it will rain, and is that probability high enough that I need to carry my umbrella?

Probability is a statement of the likelihood of an event occurring. In research, it is a statement of the likelihood that our research findings are significant (meaningful) as opposed to nonsignificant (not meaningful).

A simple example or two and two words that will become important down the road are useful here. You got up this morning and, while drinking your coffee, watched your favorite meteorologist's weather report. He or she said there was a 40% probability of rain today. That means there is a 60% chance it will *not* rain. There are our two key words: **probability** and **chance**.

Could we have said there was a 40% probability of rain and a 60% probability that it will not rain (or a 40% chance it will rain and a 60% chance it will not)? Yes, we could have said either one—but we are all budding researchers, and we do not want to do that. Here is why: Probabilities are based on data, while chance is whatever is left over.

In saying there was a 40% probability of rain, the meteorologist based this prediction on a careful statistical examination of a number of variables known to be associated with possible rain events. In other words, the meteorologist had hard data that s/he plugged into a statistical program that generated the probability. The 60%, by contrast, are all those unknowns out there that can affect weather but were not in the data. Not being meteorologists, we might speculate, nevertheless, that those data points added to the equations to get that 40% probability might include temperature, humidity, wind directions, high- and low-pressure interactions, etc.

What might constitute the 60% chance? Perhaps there are large forest fires in the southwest that were not anticipated by anyone (an unknown) and are large enough to begin exerting some influence on wind speeds and directions, and the path of lows and highs as they move around the column of hot air rising from the fires. Perhaps a nearby pulp mill cuts back on its carbon dioxide scrubber to save a few pennies (not anticipated by anybody but the management) and the chemical changes in the air modify conditions enough to affect precipitation probabilities.

The point is that probabilities are based on hard data, while chance is based on the absence of data—the unknown, the unanticipated.

Here is another example. Suppose you concoct a potion in your garage that will knock out the common cold, all 2,500+ viral variations, within six hours of onset of symptoms. You have tested this potion on mice and found it to be effective at the .05 level of probability. That means your hard data (e.g., microscopic examination of virus levels after injection, blood tests, urine tests, etc.) indicate that your potion knocks out the virus in 95 out of 100 mice. Five out of 100 mice are not cured, so chance is operating 5% of the time.

What might those chance factors be? Perhaps, unknown to you, pneumonia viruses were mixed in with some of the cold viruses, so some mice were injected with pneumonia. Perhaps, unknown to you, some of the mice had something akin to kennel cough in dogs—that is, respiratory complications from living in close quarters with other mice and their feces, dander, etc. Those chance factors were unanticipated—unknown to you—yet they affected your results, although only 5% of the time.

To beat a dead horse, probability is a statement of the meaningfulness of findings based on hard data. In this case, your potion is 95% effective—a meaningful finding by itself. Chance is a statement about the unknowns that limited the cure rate to just 95 out of 100 mice, leaving five mice uncured.

A 95% probability is good. If you had a 95% probability of winning at the craps table in Las Vegas, would you play? We sure would!

We cannot pass up the mention of Jeff Liles for a third example. He is a basketball coach and pastor who has made as many as 83 foul shots in a row without a miss—as many as 1,415 successful foul shots in one hour. Obviously, such accuracy is the product of practice and skill. When he shoots, his chance of missing is very low and the probability is very high that he will hit the shot. Given his skill level, when he misses, we can assume that chance was operating—a muscle contracted just a fraction of a bit more than it has been taught due to an ATP surge in the body, or there was a momentary distraction just at the moment of release, perhaps caused by a car backfiring or the air conditioning coming back on with a greater burst of air than normal and blowing the ball off course.

How does all of this apply to an actual research study? Let's say you are conducting a longitudinal study for the FDA on the efficacy of a new drug. You have selected your subjects and randomly assigned them to a drug versus placebo group. It is extremely important that the pill, whether actual or placebo, is taken once in the morning and again in the evening. One of your subjects gets up one morning and is about to take the morning pill when he or she hears their spouse calling from another room. They go in, find the spouse on the floor from a fall with an apparent broken leg, and rush the person to the ER. Hours later, they are back home, the pill completely forgotten in the course of the emergency. You, and your study, have just been hit by a chance event.

This was not something you could have anticipated. Whether this chance event will affect the outcome of your study is yet to be determined. Certainly, enough chance factors such as this and the outcome could be adversely affected.

Throughout this discussion of probability, we have been using the 95% mark, also expressed as a probability of .05 ($p = .05$), meaning chance is operating no more than five times out of 100. Why did we pick this level of probability? Because it is an accepted level in psychological research and in many other disciplines as well. What researchers are saying is that if their finding is influenced by chance no more than 5% of the time—if their findings are significant (meaningful) 95% of the time, they are happy.

Figures 4.1 and 4.2 illustrate this. The shaded areas in the tail are the **regions of rejection**. If a sample falls in these regions, we say that the probability of it being a "true" random sample from the population is low and it is rejected as not representative of the population. The size of the region of rejection will be a function of the probability level of the study. When $p = < .10$, the region will be relatively large as compared to $p = < .05$ or $p = < .01$. As we decrease the probability of chance being a factor, we decrease the region of rejection and increase the region of acceptance.

FIGURE 4.1 Normal Distribution Curve

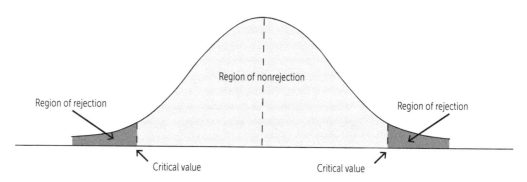

FIGURE 4.2 Skewed Distribution Curve

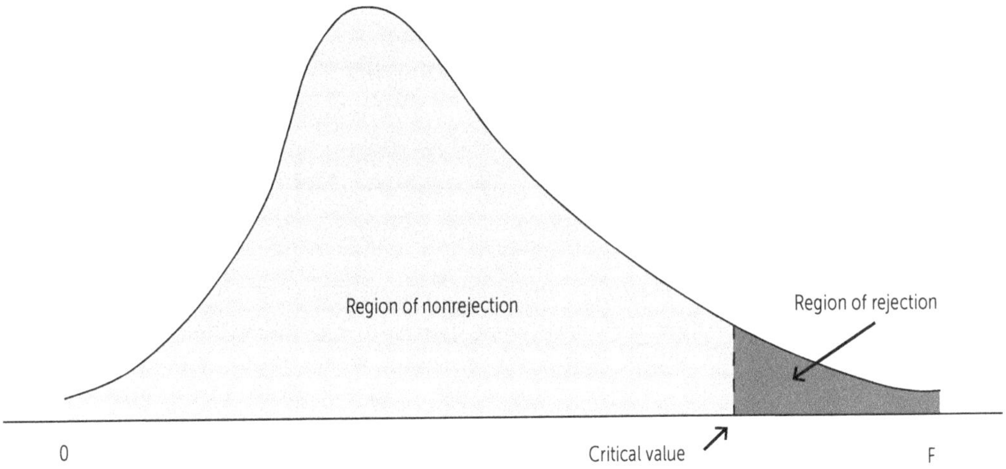

Region of nonrejection

Region of rejection

0

Critical value

F

You might well suggest that this is not very exact, and, indeed, you would be right, since there is that possibility of saying you have significant (meaningful) findings and yet being wrong 5% of the time.

Think of it this way: Would you want to live your life knowing that you would be wrong about whatever 5% of the time? Here is what we are saying: Somebody will come up to you and wave a magic wand. For the rest of your life, 95% of all decisions you make will be the right ones and 5% will be incorrect. That means 95% of your investments will be good investments, 95% of the cars you buy will be good cars, 95% of your friends will be good friends, 95% of your projects will turn out right. Would you take this good fairy up on the offer? Would a 95% probability of correct decisions put you ahead of the average person? Your authors would jump on that opportunity in a heartbeat. Where is that fairy?

There are levels of probability other than .05. Most often, you will see levels of probability such as $p = .01$ (chance operating 1 time out of 100) and $p = .001$ (chance operating 1 time out of 1,000). Occasionally, we see $p = .10$. Most often, you see this level of probability in tables giving confidence intervals based on scores, such as IQ scores. In this case, you might have a test-derived IQ score of 115, but will give the confidence interval as 106 to 121, meaning that you are 90% sure that the "true" IQ lies within this band. This confidence interval simply recognizes that, on any given day, variations (chance factors) in human behavior (how you feel, what you ate, how you slept, etc.) can influence an IQ score so that the IQ score of 115 is expected to vary along with fluctuations in the individual's current state of well-being.

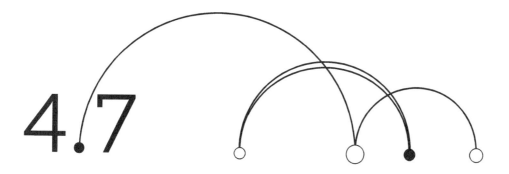

4.7

The power of randomness

We are going to go back and look at random sampling again. It is such an important means of selecting subjects and extrapolating results. We want to look at different ways of sampling in this last section.

To repeat, the primary purpose of random sampling, particularly in large numbers, is to give us a cross-section in the final **sample** (the sample are those individuals who will be part of the study) of those characteristics within the **population** (the population is the larger group of individuals from which the sample was selected) that could have an impact on whatever it is that we are measuring or observing (the measured **variable**).

We do random sampling to control **extraneous variables**. For example, if we are measuring IQ, we know that there are a number of characteristics within the population that may influence IQ and, thus, the measurement of IQ. These characteristics are referred to as **extraneous** characteristics, factors, or variables. In the case of IQ, extraneous variables might include elements such as gender, gestational time, history of illnesses or injuries over time, mother's prenatal behaviors (such as smoking or drinking, and medical monitoring of the pregnancy), richness of the growth environment, age that schooling began, drug usage (legal and/or illegal), test-taking ability, motivation, perseverance, etc.

By randomly sampling from a very large population (perhaps all the sixth-grade students in your county school district), we should end up with a sample (perhaps 200 randomly selected sixth-graders from that county school district) that has about the same percentage of males and females, premature births and full-term gestations, illnesses and injuries, etc., as the population the sample came from.

To repeat again, random sampling is the primary way we get our samples to *mimic* the population from which the samples were drawn. Mimicking the population is important if we are going to be able to generalize findings from our sample to the population.

The preferred form of random sampling is where every potential participant has the same chance of being selected each and every time a selection is made (**true random sampling**). To illustrate this, we might want to do a study of first-year college students using a sample drawn from the incoming first-year class. We might, therefore, put all the names of incoming freshmen in a basket, stir them up, and draw a name. After recording that name, we put the slip of paper with the name back in the box, stir, and draw again.

We repeat this process until we have selected our sample. If we draw the same name twice, we would just put it back in the second time and draw again (although it would be fun to write it down twice and then try to convince the subject of having to become a duplicate). The idea is that everybody has the same **probability** of their name being drawn each time.

If we write a name down and throw the slip of paper away, we actually change the probability of the remaining names being drawn, even if infinitesimally. As an example, we fan a deck of cards, with the backs of the cards toward you, and ask you to draw, blind, the 6 of clubs. Your probability is 1 in 52—.01923, or 1.923%—of blindly selecting the 6 of clubs. If you draw the 3 of diamonds and we throw that card away and ask you to try again, your probability of drawing the 6 of clubs has now improved to 1 in 51—.0196, or 1.96%. That is not a great increase in probability, but, if we continue to throw away incorrect selections, the probability of drawing the 6 of clubs will just get better and better with each draw.

There are various versions of random sampling. One is called **stratified random sampling**. That is used most often when we want to create very representative samples based on known characteristics. For example, many IQ tests are standardized nationally. Did this mean that every individual in the United States within the age range of the test was tested? Of course not—that would take a prohibitively long time. Rather, a sample was created that mimicked the population as a whole, with such samples usually defined by known parameters derived from census data.

If we know that a certain ethnic group is 30% of the population, for example, we will want to be sure that 30% of our sample comes from that ethnic group. If that ethnic group varies by region, say 20% from the Northeast and 10% from the Pacific Coast, then we will want to be sure that 20% of our sample of that ethnic group comes from the Northeast and 10% from the Pacific Coast. If 5% of the 20% of the population from the Northeast for that ethnic group comes from college-educated parents, we will want to make sure that 5% of the 20% of that ethnic group in our sample of 20% from the Northeast also comes from college-educated parents, and so on.

The idea is to duplicate the same distribution in the sample that is found in the population across important variables. This creates representativeness. It also saves time. For example, the standardization sample of the Wechsler Intelligence Test for Children, fourth edition, was only 2,200 children, but this is considered a national standardization because those 2,200 children were so carefully selected to mimic the population across important variables.

Major television networks use versions of this sampling technique when they cover elections, particularly presidential ones, and sometimes call a state for one candidate or the other based on what may seem to be a very small percentage of returns. If you know the factors that tend to influence how people vote and you know the demographics of a state, you can randomly sample how people voted within demographic groups for those factors. It generally works well, although sometimes it can backfire, as in the case of the Florida vote in the Bush-Gore presidential campaign of 2001 (the major networks declared Florida for Gore, only to find that, as more votes came in, the state was going to go to Bush).

Another version is **systematic random sampling**. As an illustration, a student was conducting a study looking at the relationship between number of friends on Facebook, according to whether a person on Facebook was single or not single. That is, do single or married individuals on Facebook report having the most friends? Since there are so many entries on Facebook, the question arose about how to randomly sample. The decision was made to take every 15th person, regardless. This is systematic random sampling, where every *n*th person is selected. The selection basis can be every *n*th person, time, event, etc.

Some sampling procedures also fall under the rubric of **convenience sampling**. When you go to the mall and someone greets you, clipboard in hand, and asks if they can ask you a few questions, you are being exposed to convenience sampling. Convenience sampling means just that: you get whatever participants happen to be handy and willing to participate. Many psychology classes have students conduct studies using classmates as participants. Unless one selects a subgroup from the class randomly, the use of the class is basically a convenience sample.

A number of years ago, one of your authors was involved in a study where data were collected on depressed children admitted to a large state psychiatric hospital. Eventually, we developed a database large enough that we could randomly sample within the database but, initially, every depressed child who came in was part of the study—because they were convenient.

One of our students is now doing a study on illicit drug usage. She has a friend who is a drug user who is participating in the study and is also putting her into contact with other drug users. Eventually, she hopes to get a large sample through this referral process. This is called **snowball sampling** because, like rolling up a snowball, it starts small but grows, providing the consistency of the snow is right. Among other factors, it means that one could hit a dead end where no participant is willing to give the names of other potential participants, a not-uncommon problem when investigating illegal areas such as drug use, prostitution, etc.

Some of these sampling procedures are very powerful, such as true random sampling or stratified random sampling. Some are not, such as convenience sampling and snowball sampling. In the case of the last two, you cannot really generalize your results to all people who go to the mall or to all people who use illegal drugs. The further away you get from true random sampling, the greater the compromise you make with your data.

Even if random sampling is not possible, random assignment to treatment conditions or groups often is. While not as ideal as random sampling, it is better than nothing. When random sampling and random assignment are used in conjunction, you have a powerful subject selection method going for you.

Where to Next?

In Chapter 5, we will look at quantitative methods. Quantitative methods—the use of numbers as your data—contrast with qualitative methods—no numbers—covered in Chapter 6.

Key Terms

- Alternative hypothesis
- Archival data
- Chance
- Confounding variable
- Construct validity
- Content validity
- Convenience sampling
- Convergent validity
- Dichotomies
- Discriminant validity

- Extraneous variable
- Face validity
- Nonprobability sample
- Null hypothesis
- Polytomies
- Population
- Probability
- Probability sample
- Random error
- Regions of rejection

- Reliability
- Representative
- Sample
- Snowball sampling
- Stratified random sampling
- Systematic random sampling
- True random sampling
- Validity

Questions

1. You walk into the mall and an individual is standing there with a clipboard who asks if he or she can ask you some questions. What type of data sampling method is this?

2. Can just the freshman class at Collier University be a population?

3. Tabulation of election results uses what kind of sampling?

4. If we set probability at .05, we are saying that chance is operating no more than:
   a. .10% of the time
   b. .01% of the time
   c. .05% of the time
   d. Less than 2% of the time but no more than 6%

5. True or false: You can have validity even if you do not have reliability.

6. You have not controlled for gender influences in your study. Gender is now what kind of a variable?

7. If I take an intelligence test at age 16, then again at age 38, and if my scores are very similar, the test has _____.

8. Once you have identified an area of interest for research, what are the values of the preliminary literature review?

9. Give two examples of dichotomies and two examples of polytomies.
   a.
   b.

10. How would you define *probability* and contrast it with chance?

11. Why engage in random sampling?

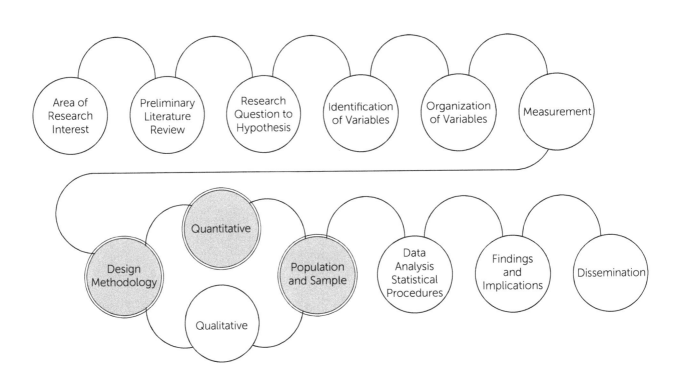

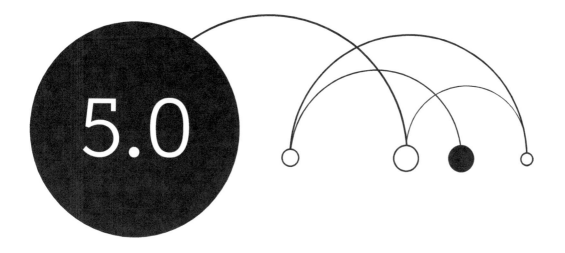

Quantitative Methods

Important research findings: 10% of all thieves are left-handed;
100% of all polar bears are left-handed; if your car is stolen,
there is a 10% chance it was nicked by a polar bear.

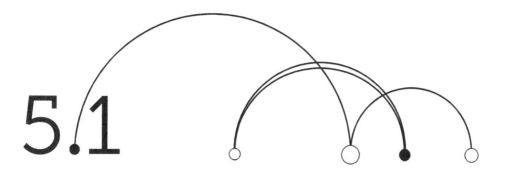

5.1

Differences between quantitative and qualitative methods

At the simplest level, **quantitative** research involves numbers that can be manipulated mathematically and/or statistically, and **qualitative** research does not. That is the crucial difference. There are, however, confusing aspects to this black-and-white definition that we will get into later. For now, just think numbers = quantitative, no numbers = qualitative.

An example may help clarify the difference. We will turn to a report from ABC World News Tonight (June 29, 2016). Why use a news report? Because so much research information is reported this way and—going back to an earlier statement—if you want to be an intelligent consumer, you must know something about evaluating research to see if it is worth paying attention to. While we would like to think all research is done well, that is not the case—some "findings" get into the public media stream that probably should not be there.

Here is the information:

> Since 1970, women are waiting 4.3 years longer to get married (average age 25.1 years), and men are waiting 3.6 years longer (average age 26.8 years).

Going back to the question of an **intelligent consumer**, can you track these findings down to the scientific study that presented them to verify their legitimacy? While you are doing that, we will get on with our discussion of these data as an illustration of quantitative research.

First, we note that the variable involved is **ratio** (age—there may also be a **nominal** variable in these data—married versus not married). We also note that means are reported, which would represent a statistical manipulation of the ratio data. While the ABC News report did not say that either men or women are waiting *significantly* longer to get married, that question should be answered, based on the data. Again, establishing significance would be a statistical manipulation. Given this, we can conclude that these findings represent a quantitative research study.

Is there the possibility of a qualitative expansion of this quantitative study? The answer would be yes, of course. You, or we, could now go out and interview men or women who have waited 3.6 or 4.3 years longer (or thereabouts) than in the past to get married and ask them "Why?" Doing so would potentially put us in the qualitative arena. It would give us the reason behind the statistics. Are men/women waiting because of pursuit of an education?—pursuit of a career?—a less-than-stellar selection of available bachelors or bachelorettes? The qualitative part can shed some light on this and give color to the rather cut-and-dried numbers.

This example illustrates two important points. The most important is that even though quantitative and qualitative are different research methods, they are not incompatible. Each can complement the other under certain circumstances.

The second point addresses that rather-confusing aspect we mentioned earlier. "Men and women" represents a nominal (dichotomous/categorical) variable. Sometimes such categories are referred to automatically as qualitative variables. While numbers cannot be assigned to the category of men and women following some mathematically dictated practice, numbers can, nevertheless, be assigned. We can make men a 1, women a 2, or vice-versa.

The assignment of numbers is, admittedly, rather arbitrary. Such assignment does not connote any mathematical properties to the assigned numbers or any hierarchy to the categories. However, once we assign numbers, those nominal variables can be used in mathematical or statistical manipulations, i.e., point-biserial correlation, discriminant function analysis, etc. The point is that nominal variables are not automatically qualitative variables. Whether they are or are not is a function of your research question (what reasons do women give for waiting longer to get married?) and how you will report the data (the primary themes of why women are waiting longer appear to fall into three areas).

With this brief example, we can already see some differences in the goals of quantitative and qualitative research—in the type of research question asked, where the data come from, the way the data are collected, and how those data are handled and reported. Figure 5.1 captures these differences and would be useful to refer back to as you read the rest of this chapter. (We will repeat the figure in Chapter 6.)

As you can see from the quantitative flowchart, the starting point is the same as suggested in Chapter 1. All research, quantitative or not, should start with an area of interest that leads to a research question of interest.

FIGURE 5.1 Quantitative Flow Chart

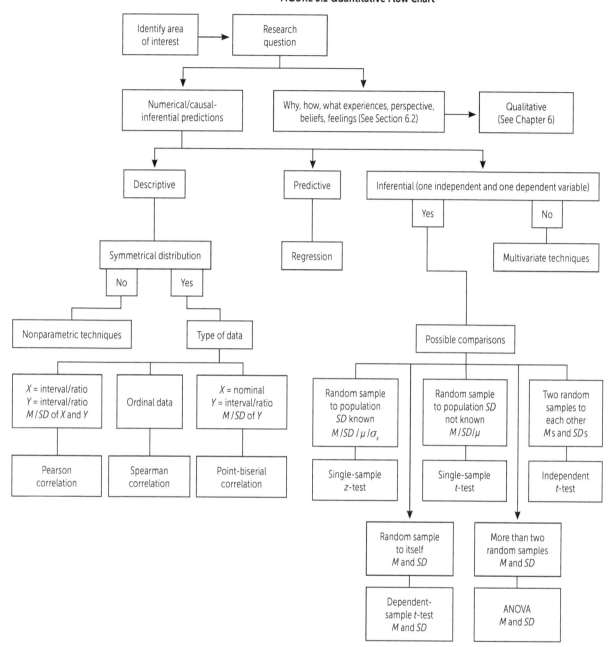

It may sound like a naïve question, but why the emphasis on "interest"? Some research projects can be long, stretching out over months or even years. For example, the dissertation research your authors completed as part of their respective doctoral programs lasted between two and seven years. That is a long time to live with a project. If it is on a topic that is of little interest, it can seem like a lifetime! The length of time involved in dissertating is also why doctoral students describe something akin to post-partum depression after defending.

Also important to note in the quantitative flowchart are the necessary decision points along the way. Will your study branch off into the qualitative area? or nonparametric analysis procedures?

Will it be descriptive or inferential? univariate or multivariate? These are important decisions that are critical to the credibility of your study. There are few more transparent aspects of a research study than selection of an inappropriate analysis technique given the research question and the nature of the data!

In Table 5.1, we summarize the differences between quantitative and qualitative research mentioned above. This summarization should be helpful as a means of keeping you on track as we talk about each research method and the similarities and differences between the two.

TABLE 5.1 Comparison between Qualitative and Quantitative Research

Criteria	Quantitative	Qualitative
Objective	Quantify and generalize data from sample to population	Develop understanding of underlying reasons and motivations
Research question	What, why, how (differences) Relate, compare	What, why, how Describe, explore, understand
Sample	Large, random	Small, nonrandom
Data	Numbers	Words
Analysis	Statistical	Nonstatistical
Sources	Measurements and objective instruments	Interviews, observations, etc.

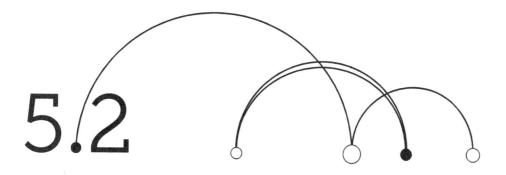

5.2

Essential characteristics of quantitative methods

As already suggested, quantitative research uses numbers. Other essential characteristics center around the goals of the research, the research question, data sources, handling the data, and generalization of findings. We will take each one separately. Then we will look at limitations of quantitative research. Everything has limitations, does it not?

Goals of Quantitative Research

Quantitative research, at its heart, is an attempt to expand the boundaries of knowledge in a systematic, predictable, and verifiable fashion using numbers. The methodical and controlled nature of the data gathering and data analysis(es) is aimed at eventually being able to extrapolate findings from a small body of participants (the sample) to a larger body (the population). Such extrapolation depends on the rather rigid adherence to a set of requirements that maximize both the extrapolation, and the power of the extrapolation.

Basically, this means that the entire procedure from beginning to end is reasonably inflexible, prescribed, guided by pre-determined parameters, and subject to objective scrutinizing by others to validate that prescribed procedures were followed (quantitative research, given its rigidity, is a field in which a healthy dose of obsessive-compulsiveness is an adaptive asset). The highly controlled methods of quantitative research are what allow us to edge forward, slowly expanding the boundaries of knowledge in a systematic, step-by-step manner.

The Research Question

Research questions in quantitative methods are focused, neutral (no pre-supposed outcome implied by the question, usually), simple (usually), parsimonious (usually), and imply the use of analytical systems grounded in math and/or statistics (such as saying there will be no significant differences between A and B) to answer the research question at a specified level of confidence (probability) while minimizing chance and controlling for extraneous variables and, thus, alternate explanations (usually).

To boil down that rather long sentence, the research question is one that, by its very nature, must be answered empirically. In other words, any variable involved in the measurement must be **operationalized**—defined in a manner that allows quantification. If you wish to study constructs such as love, anger, depression, intelligence, etc., you must get them to a measurable level. For example, you could use a well-known intelligence test to quantify intelligence, the Beck Depression Inventory to quantify depression, or the Children's Inventory of Anger to quantify anger. Each of these instruments is broadly accepted, has adequate reliability and validity, and yields numbers for the constructs.

Research questions may be stated as **null hypotheses** (no difference) or as **alternative hypotheses** (statement of the difference expected). The questions generally try to describe a relationship, establish causality between variables/events, or predict some phenomenon.

To illustrate, here is a research question stated as the alternative hypothesis: Are women older now when they get married as compared to 2006? The null hypothesis would be: There is no difference in the average age of marriage for women in 2006 as compared to now. This would be a descriptive research question since it simply asks if a difference exists, not why.

We can change it to a causal question as follows: Why has the average age of women getting married increased as compared to 2006 (assuming that we have already established that they are waiting longer). This research question can be addressed quantitatively or qualitatively—your choice. You can interview women (qualitative) or you can collect additional quantitative data (perhaps through a life-satisfaction questionnaire) that can be subjected to analyses that provide reasons why women are waiting longer.

Data Collection Techniques

For quantitative research, data are usually collected using well-known techniques (for example, GSR—Galvanic Skin Response) or measures (such as the Armstrong Laboratory Aviation Personality Survey; ALAPS) that have established **reliability** and **validity**. For example, individuals undergoing a trial study in the use of Transcranial Magnetic Stimulation (TMS) may respond to a depression inventory every two weeks during the normal six-week procedure to determine whether progress is being made. The depression measure would probably be a known and valid measure, such as the Beck Depression Inventory.

Whenever possible, participants are selected by a **random selection** process to ensure that the sample mimics the general characteristics at play in the population. This is the primary method of controlling extraneous variables such as ethnicity, gender, etc.

Let us suppose we wanted to do a study on the Citadel Corps of Cadets—a total of some 2,300 potential participants. We randomly select a subset of 200 cadets. This is our sample. Now suppose that the sample consists of 12% female cadets, where the base rate in the Corps is only

5%. Obviously, females are overrepresented in the sample as compared to the population (the Corps) and, therefore, gender is an extraneous, or uncontrolled, variable in the sample.

Data collection techniques for quantitative research have to be **transparent** (except where deception is necessary and justified), **systematic**, **verifiable**, **replicable**, and **objective**. While each of those criteria makes sense, systematic is a bit tricky. For example, if height is an important variable in your research, height varies throughout the day: People are taller in the morning than they are in the evening. We lose fluids in our joints, tension in our muscles, and motivation in our stance, thus leading to slumping, slipping, and squishing. If height is important and you wish to be systematic, measure height either first thing in the morning or last thing in the evening—and at the same time of day for all subjects.

Data Sources

In quantitative research, data come from surveys, questionnaires, measurements using established instruments (usually), manipulation of events (experimental conditions), or archival sources. Importantly, whatever the source, the data are handled as an aggregate—there are no attempts to identify individuals or present individual data except, perhaps, as postscripts to the research to illustrate findings. Even then, any individual identification, even if by affiliation with an institution, etc., must be protected. If such postscripts can lead to identification of particular participants or groups of participants or institutions, they are to be avoided unless permission is received (see Section 9.2 for further clarification of this requirement).

Data can come from direct or indirect observation, although this tends to be more the bailiwick of qualitative research. For example, we might monitor airplane instruments for pilots-in-training to assess accuracy of touchdown on a runway. Across a number of student pilots, those data can be aggregated and used to evaluate the training program (quantitative relative to such accuracy). Conversely, such data can be used in individual feedback to students to direct them to remedial training or advancement, etc. (qualitative).

Data Handling

In quantitative research, data are usually entered into computer-assisted analysis programs such as **SPSS** (Statistical Package for the Social Sciences), **SAS** (Statistical Analysis Software), or myriad other programs (type "statistical analysis programs" into your search engine and see how many pop up). Of course, using a statistical package is not necessary for quantitative methods. Some analyses are easy enough to do on a calculator (e.g., correlations). In most cases, however, the statistical package will save time and effort, and minimize errors—if the data are entered correctly.

Generalization

Generalization is the heart and soul of quantitative research. Quantitative research is a broad paintbrush that gives us variable associations and relationships, and comparative data across groups and/or manipulations that are statistically significant. It is the counter to idiographic anecdotes, personal experiences, individual interpretations, singular opinions, stereotypes, and outdated axioms and prejudices.

An example is in order. The **Flynn Effect** describes a broad trend throughout westernized industrialized countries that is grounded in numerical data, examined statistically. It is interesting, has some significant implications, and may say a good deal about the direction we are going. So you ask, just what is this Flynn Effect? We are not going to tell you. The reference is at the end of this chapter—you can actually listen to Professor Emeritus James Flynn explain his own work on TED.com (Technology, Entertainment, and Design—type in "Flynn Effect"). He is an erudite and entertaining speaker—and the session is only about 15 minutes long.

Limitations of Quantitative Research

One of the primary limitations of quantitative research is that it is just that: quantitative. It is all about broad trends and numbers and, as such, may lose important considerations such as motivations of participants, participant attitudes and perceptions, and the essence of what it is like to be human. Even if data collection instruments are designed to tap motivations, etc., pre-arranged answers may not be broad enough to capture participant responses, thus channeling those responses into the particular answer set constructed by the researcher and creating potential researcher bias. This would be true of any survey or questionnaire developed by the researcher.

In addition, the relatively inflexible nature of quantitative research dictates that, once a course is set, it must be followed. It is very difficult to go back and build in questions to gather additional information you discover you need but did not realize was important until after the project was underway.

The scope of quantitative research must, by its nature, be somewhat limited. Remember that two of the key characteristics of a good research question are that it is simple and parsimonious. Elaborate, involved research questions may capture the complexity of a particular area of research but may be too complicated to answer using statistical methodology, or the methodology may produce results too complex to be broken down into smaller components that are actually understood (George & Mallery, 2012).

The limited scope of quantitative research questions might also mean that you miss the trees for the leaves (or is that the other way around?). You know what we mean: The broad picture may be hiding behind the details of the data, perhaps unseen and unnoticed, and therefore missed.

The last major limitation is that behavior does not occur in a vacuum—it has contextual determinants that may be social, economic, political, etc., in nature. Quantitative research may not adequately address that context. This is where qualitative follow-up may come in. If we look at **Solomon Asch**'s famous studies on conformity, we can see this in operation. Asch had naïve subjects judge line lengths in a group situation where the other participants were confederates of the researcher and directed to give wrong answers. Thirty-five percent of the naïve subjects went along with the incorrect group choice. To find out why, Asch then went back and interviewed that 35%.

Based on those interviews, he uncovered some of the social pressures that were operating on the naïve subjects (they wanted to be accepted by the group, they did not want to make waves, etc.; Schwartz, 1986). This is an excellent example of quantitative research being followed by qualitative methods to clarify the contextual pressure to conform to group opinions—to add color to the numbers, if you will.

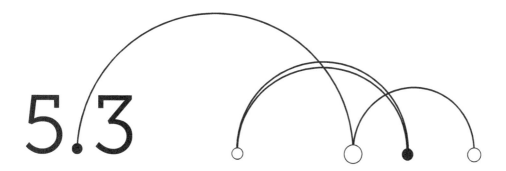

5.3

Quantitative methods: Timeframes

When you think about broad findings, broad trends, or the effects of interventions over time, you naturally also tend to think in broad timeframes. Timeframes of more than just the here-and-now can be handled by quantitative or qualitative methods. In quantitative research, there are generally two ways to deal with a long-term view that go beyond the here-and-now. One is with a **longitudinal study**, and the other is using a **cross-sectional study**.

Longitudinal versus Cross-sectional Research

In commercial aviation, there is something known as the Age 60 rule that requires commercial plots to retire at age 60 (in some foreign countries, pilots can fly past age 60 if the copilot is age 60 or less). The concern is the potential increase, at age 60 and above, of sudden medical incapacitation such as cerebral infarction, cardiac arrest, etc. With a plane full of perhaps several hundred people traveling from hither to yon, such a medical emergency can have devastating consequences.

Let us suppose that you are an aviation researcher and you want to study the Age 60 rule. Specifically, you want to collect data that will give some insight into the efficacy of the requirement—does retiring at age 60 really reduce medical emergencies on the flight deck? Perhaps you find a large number of pilots in their 20's who are willing to participate in your study. You decide to follow them throughout their careers, gathering medical and flight physicals, illnesses, stressors, etc., as part of this study. The idea is to look by decade to see if there are any particular competency trends or changes in your sample due to stress-induced CVAs (cardio-vascular accidents) or other medical emergencies.

Off you go—for the next 35 to 40 years! While you might gather and find some interesting things along the way, you will not get to the Age 60 question until your subjects are at age 60, at which time you also may be at age 60—or beyond—and ready to retire yourself.

What we just described is a longitudinal study: following a group of individuals across a span of time to see what happens over that time.

There have been a number of such studies. Perhaps best known is **Lewis Terman**'s longitudinal study of the gifted that started in 1921 and ran until 1955, then was picked up by **Robert R. Sears** and ran until 1986. That constitutes two researchers' lifetimes! Another example would be the longitudinal study of temperament by **Thomas, Chess, and Birch (1968)** over 12 years. Longitudinal studies can last for long periods!

Longitudinal studies are generally observational studies in that the subjects are not manipulated or required to do anything outside what they might normally do. Observational, in this case, does not necessarily imply qualitative. The "observations" can be measurements or naturally occurring events at particular points. For example, one could follow individuals with high cholesterol over a specified period of time, recording their cholesterol levels.

One can also conduct longitudinal research that involves some manipulation. For example, programs have been tried that involved incarcerated individuals training service dogs. The program may be new—thus, a manipulation—but once instituted with training accomplished, it is allowed to go on of its own accord and the impact on prisoner behavior is observed.

Suppose you do not want to grow old following your 20-year-old pilots until they are 60? You could do a cross-sectional study. This would mean, perhaps, having a sample of pilots who are in their 20s, another sample in their 30s, and another in their 40s, etc. Now you can measure each age cohort on the same instruments, or compare their medical/flight physicals across age cohorts. You essentially end up at the same place, but in a much shorter timeframe. Again, this is an observational approach—no manipulations and only collection of data that would normally be available.

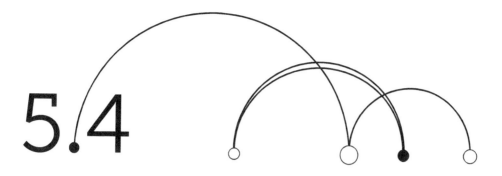

The broad types of quantitative methods

Quantitative methods cover a broad spectrum and can generally be divided into **descriptive** and **inferential** (causal) methods, **parametric** and **nonparametric** methods, and **univariate** and **multivariate** methods. We use the term "methods" here to designate general classes of statistical techniques (see Figure 5.1). To break this down, let us first look at descriptive and inferential methods.

Descriptive methods simply generate numbers that describe. Going back to our "waiting to get married" data, the average ages for marriages today are simply descriptions of what is currently going on. Additional methods can look at the "why"—these would be inferential/causal. That is, they would attempt to discover reasons why individuals are waiting later or to identify factors that are influencing that decision.

Generally speaking, new areas of research begin with descriptive methods in an attempt to pin down what a particular field looks like and what variables come to the surface, descriptively, that seem to be important in the new area of inquiry—essentially, getting the lay of the land. Once the landscape is known to a greater extent, one can ask why the landscape is as it is—inferential/causal questions.

A simple example: You buy a large tract of land to start growing something for sale in the market. When you first see the land, you note that it is relatively flat (descriptive), gets an average of 32 inches of rain per growing season (descriptive), and is in a region where about 80% of the days are sunny to partially sunny (descriptive). You start growing corn.

After the first season, you think, based on past experience, that your yield is low. Now the question is "why?" (inferential/causal). You begin taking random soil samples for analysis and, based on results, start adding specific minerals to the tract of land (your manipulation). After the second growing season, you compare yields and find that the current yield is significantly greater than last year's yield, which you attribute to the use of the minerals applied before the second growing season (causal).

In addition to descriptive and inferential methods, univariate and multivariate methods are available in quantitative research. You ask, of course, what exactly is univariate and multivariate?

Univariate research focuses on only one variable—height, as mentioned in the importance of being systematic in collecting data; age, in the case of our ABC data; or the relationship between two variables, such as IQ and grade point average. Univariate techniques might include **correlations**, *t*-tests, **Analysis of Variance (ANOVA)**, etc.

The point is, univariate methods are singular in nature. Multivariate is the opposite, as "multi" might imply. Multivariate methods are in response to the recognition that rarely does one dependent variable or one manipulation describe the complexity of human behavior or variable and context interaction (Harris, 1985). Multivariate techniques might include multiple regression, factor analysis, etc. Multivariate techniques begin to address one of the limitations mentioned earlier—that of the narrow scope of numerical analysis. Both univariate and multivariate methods come in descriptive and inferential flavors.

As if descriptive, inferential, univariate, and multivariate were not enough, we also have **parametric** and nonparametric research methods. Very simply, parametric methods are based on the assumption that the data approximates a symmetrical (near normal) distribution. Not all samples do approach symmetry, but there are ways to determine if they are close enough (**skew** and **kurtosis**).

In addition, parametric methods are quite robust and can tolerate some give in the assumption of symmetry. Nonparametric data make no assumptions about the symmetry of the distribution. Nonparametric methods are used with small sample sizes, skewed samples, etc.

We can have any combination of these parameters; for example, descriptive-univariate-parametric methods, or inferential-multivariate-nonparametric methods. Generally speaking, there are separate, higher-level statistics courses for each—a course in univariate methods, another in nonparametric methods, and yet another in multivariate methods.

Some Basic Quantitative Designs

At the core of quantitative methods is the **experimental method** or **experimental design**. We alluded to this when we presented the example above, dealing with your tract of land, although we need to elaborate on that a bit. Speaking of tracts of land, experimental designs grew out of agricultural experimentation and beer-brewing endeavors (see, research methods *are* relevant!). The basic designs were developed by **W. S. Gosset** (known as Student), who worked at Guinness Brewery, and Sir Ronald Fisher, who worked at Rothasmsted Experimental Station (Tankard, 1984). All of this was in the early 1900s.

As an aside, most univariate techniques we use today were developed during this period. We have only to add **Sir Francis Galton** (**Charles Darwin**'s cousin) and **Carl (Karl) Pearson** to round out the list of early pioneers (Tankard, 1984).

Back to experimental designs. This is perhaps best captured by Fisher, who selected multiple growing fields in very close proximity so that rainfall, sunshine, soil composition, etc., were equivalent. Corn, or another crop, would be planted on each field and then a manipulation, such as different types of fertilizers, etc., would be applied to the different fields. The final yield would be tabulated and compared across the fields and fertilizers to identify the fertilizer that produced the greatest (significant?) yield. That is it in a nutshell.

A similar study today might involve pilot training, where students are **randomly assigned** to three groups for training purposes—a group that does not use simulators, a group that does use simulators, and a third group that watches control manipulation on a video. Relevant skills would be measured across the groups and the data examined for significant differences in the training methods. The goal would be to determine which method worked best: no hands-on or visuals, hands-on with visuals, or visuals only.

The key characteristics are, ideally, randomly selected individuals randomly assigned to different experimental (treatment) conditions, one of which can be a no-treatment control, where there are a minimum of two such conditions (which would be t-tests) or more than two (which would be ANOVA). Under less-than-ideal circumstances, one may not be able to select subjects randomly. As in our example above, you cannot randomly admit individuals into pilot training—that is an individual prerogative. However, random assignment to conditions should not be a negotiable part of experimental methodology if extrapolation of findings is a goal.

Another quantitative method is the quasi-experimental, or design. It is exactly like the experimental method just described, but with one important difference—the variable under study is not open to random assignment.

Let us suppose that you work for the Federal Drug Administration (FDA) and you are involved in testing new drugs. Creative Assorted Pharmaceuticals (CrAP) has developed a new pill for the treatment of depression. You are in charge of setting up the design to look at the efficacy of this new drug.

You decide you will have three conditions: a placebo, an SSRI condition, and a new drug condition. You cannot go out and randomly select subjects who may or may not be depressed and press them into your study, suggesting that those who are not depressed need to become so quickly. You can, however, identify individuals already suffering from depression and randomly assign them to one of the conditions. Depression is a quasi-experimental variable—it is usually endogenous to the individual and cannot be randomly assigned due to the negative, harmful, or dangerous (suicide?) characteristics of the variable.

There are many quasi-experimental variables, such as smoking, emotional disturbance, Alzheimer's, etc. You can study such variables, but only with those already manifesting the variable.

Correlation, as another quantitative method, can be attributed conceptually to Sir Francis Galton, Charles Darwin's half-cousin. **J. D. Hamilton Dickson** of the Cambridge math department worked out the initial math for the method and it was later refined by Carl Pearson (Tankard, 1984). The technique examines the potential for relationships between variables that intuitively might seem to be related, such as intelligence and grades, fluctuations in the length of women's hemlines and stock market activity (yes, there is a relationship; see Mabry, 1971), cow flatulence and atmospheric methane levels (Drennen & Chapman, 1992), etc. Correlational methods simply identify a mathematical relationship. They do not, in any way, establish causal factors.

Last in the list of quantitative designs are **survey or questionnaire** methods. Survey methods use surveys to gather data. (That statement was probably obvious!) Surveys and questionnaires are generally a paper-and-pencil method of gathering data by asking questions and having subjects respond to the questions by choosing from some set of offered choices, such as "strongly agree" to "strongly disagree" or forcing subjects to choose between two alternatives, as in the Myers Briggs Type indicator (open-ended responses would fit more with qualitative methods). There is a proliferation of surveys/questionnaires on the World Wide Web; one has only to visit sites such as www.surveymonkey.com to see a number of different survey-type instruments in a broad range of topics.

By their nature, survey methods seem like they would be easy to do—just construct a set of questions with some response scale that yields the equivalent of a score on the survey. In reality, surveys—*good* surveys—are difficult to construct and require a clear notion of what the researcher is after, the expertise to know what type of questions should be asked, and a good understanding of the mechanics of constructing a survey.

Relative to that last point, and simply as an indication of what you could be up against, the mechanics include considerations such as the difficulty level of the vocabulary used or whether jargon should be used, reverse-scored questions, overly complex questions, questions that are not clear and subject to misinterpretations, sexist or racist implications in the questions, privacy issues, legal issues relative to the behavior being investigated … and on it goes. It takes a certain skill set to construct good surveys or questionnaires, and not every such instrument displays that skill set. Again, look critically at surveys/questionnaires you might find online.

Where to Next?

We have just finished one of the two major branches of research methods: quantitative methods. As you have learned, quantitative methods deal with numbers; however, what we have next for you is even more exciting. In the next chapter, we explore *qualitative* research methods. Qualitative methods use works and nonstatistical analysis to look at a problem. Qualitative methods allow us to develop an understanding of the underlying reasons something is how it is.

Turn the page, and let's get going.

Key Terms

- Alternative hypothesis
- ANOVA
- Carl (Karl) Pearson
- Charles Darwin
- Correlation
- Cross-sectional
- Descriptive
- Experimental methods
- Flynn Effect
- Inferential statistical techniques
- Intelligent consumer
- Kurtosis
- Longitudinal

- Multivariate
- Nominal variable
- Nonparametric
- Null hypothesis
- Objective
- Operationalization
- Parametric
- Qualitative
- Quantitative
- Questionnaire
- Randomly selected
- Ratio variable
- Reliability
- Replicable

- SAS
- Sir Francis Galton
- Skew
- Solomon Asch
- SPSS
- Survey
- Systematic random sampling
- *t*-test
- Transparent
- Univariate
- Validity
- Verifiable
- W. S. Gosset

Questions

1. How do univariate and multivariate techniques differ?

2. How do parametric and nonparametric techniques differ?

3. Write a descriptive research question.

4. Write an inferential/causal research question.

5. Write an alternative hypothesis, then the null hypothesis for the alternative hypothesis.

6. Operationalize the following research question: There will be no difference for space required for female commercial passengers compared to male commercial passengers.

7. What are at least two potential limitations of quantitative research?

8. Give and explain at least two essential characteristics of quantitative research.

9. Can you have a study in which quantitative and qualitative methods are combined? Why or why not?

10. Why use random selection?

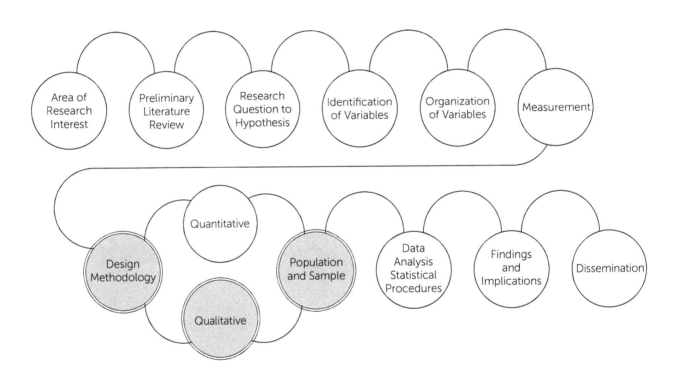

Area of Research Interest

Preliminary Literature Review

Research Question to Hypothesis

Identification of Variables

Organization of Variables

Measurement

Design Methodology

Quantitative

Qualitative

Population and Sample

Data Analysis Statistical Procedures

Findings and Implications

Dissemination

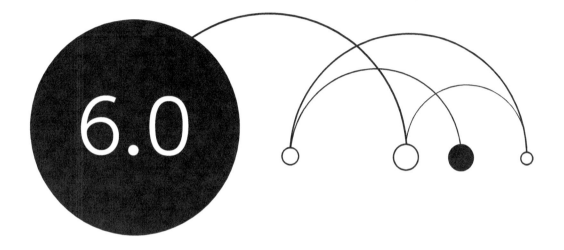

Qualitative Methods

A student in a research methods course did not study for the exam. It was a true-false test, so he flipped a coin for the answers. The professor watched the student as he flipped the coin and wrote down the answers. At the end of two hours, everyone had left except for this student. The professor walked up to the student and said, "I have been watching you flip a coin to get answers. Obviously, you did not study for the test. What I don't understand is why you are still here if all you are doing is flipping a coin." The student replies, rather irritated, "Shhh! I am checking my answers."

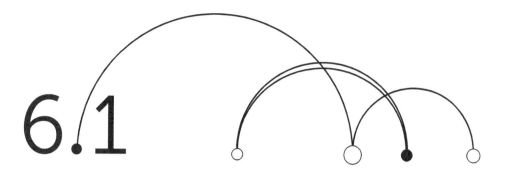

6.1

Differences between qualitative and quantitative methods

Did we not just cover this in Chapter 5, you are asking. Yes, we did, when we suggested that the main difference between qualitative and quantitative research methods was the absence (qualitative) or presence (quantitative) of numbers. Now we need to expand on that some.

In Chapter 5, we suggested that **quantitative methods** were involved with numbers and were methodical, controlled, reasonably inflexible, and guided by pre-determined parameters and a great field for people with a healthy dose of obsessive-compulsiveness. We want to remove, however, any thought that qualitative methods are the polar opposite and a great field for people with a *laissez-faire* approach to scientific inquiry. Indeed, the comparatively greater lack of structure of qualitative research methods demands paying even more attention to process and procedure—and, thus, structure—than with quantitative methods.

Part of this greater need for attention in qualitative methods is a function of the use of words over numbers. Numbers are clear and generally lack ambiguity. Anyone who makes a living with words is clearly aware of this and often skilled in using words to maximize obfuscation and minimize clarity. In qualitative research, however, obfuscation or lack of clarity, whether intentional or not, is not considered a useful approach.

Additionally, qualitative methods allow for a greater degree of ingenuity/creativity in constructing designs and analysis methods. In quantitative research, you cannot, for example, apply your creativity and ingenuity to the revamping of a Pearson Correlation Coefficient formula. That formula has been set since 1896 in Pearson's paper, *Regression, heredity, and panmixia* (Tankard, 1984) and none of us is likely to have the mathematical genius to improve on it. You can, however, create new ways to code, categorize, label, or recombine your qualitative data or qualitative data from another study (see Figure 6.1, Common Data Analysis Methods). Since you can be creative, it is up to you to be transparent and careful in what you are doing.

Another difference is the complementary relationship of quantitative and qualitative methods. We used an example of this early on in Chapter 5. To expand on that, we can now suggest that qualitative methods might be the most useful initial approach to a new area of inquiry and research. Qualitative inquiry may well be the best approach for identifying variables initially that may be subjected to quantitative examination later. You can think of qualitative research as a stand-on-its own method or as potential bookends to good quantitative research. The latter is called a **mixed-methods** approach.

Additionally, while quantitative research is more concerned with the amount of something or the magnitude of something (e.g., the size of the difference), what are referred to as closed-end research questions, qualitative research is more focused on the **what, how, why** of some event or phenomenon, a more open-ended approach (Mayer, 2015). The what, how, or why does not have to be couched in terms of the magnitude. It simply has to be described. This difference tends to make quantitative methods more deductive (from theory to findings) while qualitative tends to be more inductive (from findings to theory) (Graue, 2015; Mayer, 2015).

The important point here, however, is that when you reduce external controls, such as those in play in quantitative research, the responsibility increases for the researcher to make sure the qualitative research stands as good research.

We repeat the table from Chapter 5 (see Table 6.1), which condenses comparative points between quantitative and qualitative research methods, for continuity's sake and for an optical boost (that's like a visual!).

TABLE 6.1 Comparison between Quantitative and Qualitative Research

Criteria	Quantitative	Qualitative
Objective	Quantify and generalize data from sample to population	Develop understanding of underlying reasons and motivations
Research question	What, why, how (differences) Relate, compare	What, why, how Describe, explore, understand
Sample	Large, random	Small, nonrandom
Data	Numbers	Words
Analysis	Statistical	Nonstatistical
Sources	Measurements and objective instruments	Interviews, observations, etc.

We also present a qualitative flowchart here, as a parallel to the quantitative flowchart found in Chapter 5 and above (see Figure 6.1). You will note that the top part of this flowchart is the same as the top part of the quantitative flowchart, which is the same as the brief flowchart presented in Section 1.6. Identifying an area of research interest and formulating a research question constitute the driving force behind all research, whether quantitative or qualitative.

FIGURE 6.1 Qualitative Methods Flowchart

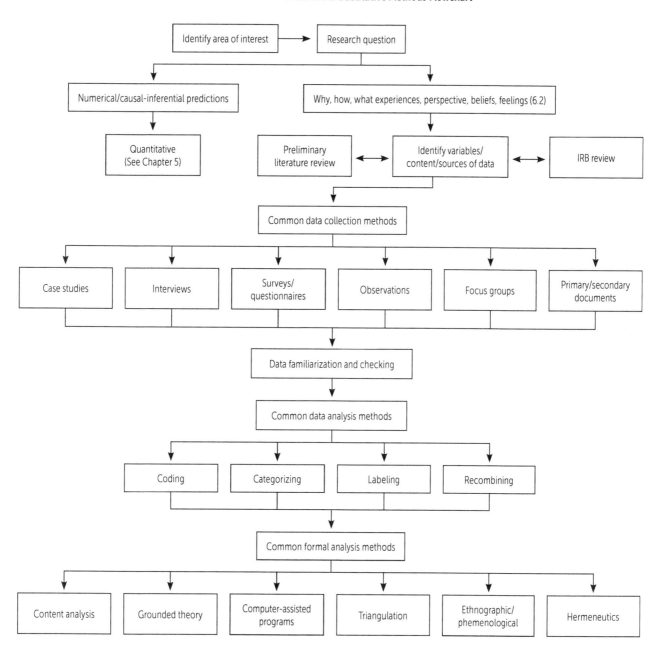

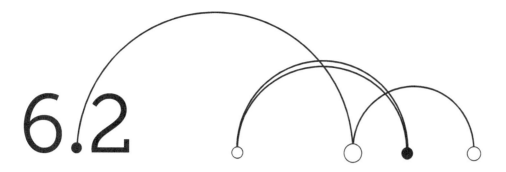

6.2

Essential characteristics of qualitative methods

An essential characteristic of both qualitative and quantitative methods is that the research is carried out in a rigorous, transparent, objective manner so the results are **trustworthy**. As we have noted, achieving this goal may be more difficult in qualitative research, due to the lack of a rigid structure that defines the process of research, particularly data analysis. It is more difficult to find a well-conducted independent *t*-test lacking in trustworthiness than to find an opinion poll lacking in trustworthiness, particularly if the opinions run counter to one's own. We can go back and emphasize that statement about *rigor, transparency, and objectivity equals trustworthiness, since it is crucial to all research* but has to be attended to even more assiduously with qualitative research.

Qualitative research has, as a central goal, the aim of answering the questions of what, why, and how as they relate to some event, group, or phenomenon without the use of numbers. If quantitative methods seek **objective knowledge**, qualitative methods seek knowledge that is more **subjective**. While it might be appropriate to understand, objectively, that there may be differences between Republicans and Democrats on the Conservative-Liberal continuum, it might be just as informative—and as necessary for working together across the aisle—to understand the hows and whys of those differences historically, philosophically, and within the current social, economic, and political context.

We also should throw religion into that mix. Why throw in religion? Simply to illustrate that qualitative methods can cut across, integrate, elucidate, and encourage discourse along multiple dimensions simultaneously. It is not as rigidly bound to a particular discipline, as a quantitative psychology study would be, or a quantitative eco-biological study, etc. In a quantitative study of Republicans and Democrats, quantitative findings just tell us if there is a difference. The more-subjective qualitative data may give clues to the multidimensionality of the constraints created by those differences.

Goals of Qualitative Research

In addition to the how, what, and why, a primary goal of qualitative methods is to generate concepts, constructs, and theories about an event, group, or phenomenon under investigation: the inductive approach. An excellent example of this would be **Freudian theory**, developed almost exclusively from Sigmund Freud's own self-analysis along with his interactions with patients, particularly such influential ones as Little Hans (Herbert Graf), Wolf Man (Dr. Sergei Pankejeff), and Anna O. We should clarify that Freud never directly treated Anna O.—Bertha Pappenheim. She was a patient of a colleague of Freud's, Dr. Joseph Breuer, whom Freud consulted with on the case.

If you are familiar with Freudian theory, you know that it is a theory of words, not numbers, where those words are crafted carefully to explain a large range of observable behaviors.

Another goal of qualitative methods is to supply a more-holistic take on some event, group, or phenomenon. While numbers, as mentioned, can be very narrow in perspective, qualitative research can be much broader and encompassing, with integration within and across dimensions or disciplines to form a larger gestalt (an organized whole that is perceived as more than the sum of its parts). Again, to use Freud's theory as an example, it encompasses, integrates, and provides a gestalt for all of the neuroses and, through dialectic contrast, normal development and behavior. While Freud's theory has not lent itself to operationalization and intensive quantitative scrutiny, it is still hard to think of the fabric of this modern era sans Freud.

Yet another goal of qualitative research is to uncover underlying contributors to behavior—more exactly, opinions, beliefs, motivations, feelings, experiences, contexts, and perspectives. These more-subjective aspects of the human experience are valuable for understanding behavior and, while they can be identified through quantitative means, qualitative inquiry may be more straightforward. It is easier to say "How did you feel about that?" than to get the same information from a 20-question instrument.

If we could wrap the goals of qualitative research into one encompassing statement, it would probably take us back to our paintbrush analogy. We would say the goal of qualitative research is to add broad brushstrokes of color either before, after, in conjunction with, or in lieu of the black and white of numbers.

The Research Question

Very quickly, and to beat a dead horse, the research questions in qualitative methods deal with what, why, and how. For example, given the current healthcare debate, a good qualitative research question might be: What do lower-income individuals prefer to see in a healthcare bill? As another example, we might ask residents of a high-crime area how they feel about the crime rate there.

The focus of qualitative research questions can be directed toward individuals, groups, the exploration of some event or phenomenon, or the observation of some event or phenomenon. Research questions that might go with each of these would be:

- How do working mothers feel about the current maternity leave guidelines in the United States? (individual and/or group responses)
- What were the driving forces behind the social and political upheavals of the 1960s? (exploration of a phenomenon)
- How will crowds behave at the Boston Marathon after the 4/15/2013 bombing? (observation)

At the heart of each of these research questions are the subjective perceptions of the participants through either expressed opinions, perspectives, feelings, experiences, future/past expectations and interpretations, or behaviors. In some of these examples, data might consist of individual answers that probably would be categorized (see Figure 6.1) based on some rubric and reported as aggregate findings to safeguard individual identities.

We stated earlier that quantitative research questions ideally should be objective, neutral, operationalizable, simple, and parsimonious. Qualitative research questions should be the same—with the exception of operationalizable. Ambiguous, unfocused, or throw-in-everything-along-with-the-kitchen-sink questions should be avoided.

Common Data Collection Methods

Data collection techniques (see Figure 6.1) will be driven by the research question. At the most general level, data collection can be direct or indirect. Asking mothers how they feel about maternity leave guidelines might be very direct. Observing to see if the bombing at the 2013 Boston Marathon has changed subsequent behavior might be much more indirect.

A common technique is the individual interview. In the realm of research, "interview" means more than just sitting down and winging it on questions. It is a sophisticated process that requires considerable thought. We will not get into interviewing theory and practice in detail here, but recommend Edwards and Holland (2013) for the serious student of interviewing. We will mention just a few considerations to give some idea of the potential complexity of the process.

At a general level, interviews can be thought of as existing along a continuum of structure from highly structured to open-ended and everything between. They generally have a **theme**—a topic—around which the interview revolves. By their nature, they are **dialogue**-driven and **transactional**. Transactional simply means that the interaction shapes both the process and the actors in the interaction. The degree of the transactional nature of the interview will be related to the degree of structure—the greater the structure, the less influential the transactional process in shaping the exchange.

The interview tends to view knowledge as contextual, hence the need to approach the source of the information—the interviewee—directly. If you were to conduct a qualitative inquiry into potential solutions to some geopolitical issue, you might interview heads of states as individuals who are intimately acquainted with the issues involved and able to speak knowledgeably about those issues—they are the source, and their perspectives are shaped by the context they come from; Asia versus the Middle East versus Europe, etc.

Beyond these general considerations are ones of the credibility of the interview. Quantitative considerations analogous to credibility would be reliability and validity. In terms of credibility, one would look at the interviewer, interviewee, and interview technique and ask whether they are objective, transparent, reproducible, and trustworthy. If the answers to these questions are yes, then the interview will appear to have credibility (validity). The skill of the interviewer, and the interviewer's objectivity and neutrality, are critical issues in the credibility (reliability and validity) of the information obtained.

Beyond these general considerations, there are the nitty-gritty elements that must be considered. Where will you conduct the interview? Can the place exert an influence (called *demand characteristics*—think of the Milgram studies at Yale and what influence this prestigious institution had on subjects' responses)? Will you have a statement dealing with informed consent and withdrawal from the interview without prejudice? Will you start with a statement that orients the person being interviewed to the topic area or will you simply start with questions?

If you are asking sensitive questions, where will you put those—up front or at the end of the interview, when perhaps some rapport has been established through the process itself? Will you have a set of questions that you follow closely, every interviewee getting the same set in the same order? Will the questions be constructed to produce yes/no answers, very short answers, or open-ended answers? Will you use prompts to expand on answers given? Will you ask questions with two questions imbedded in one? Will you have mechanisms built in to return to the central theme if the interview gets off track? Are there cultural considerations that might influence the pragmatics of the interview, such as sitting facing each other or positioning more obliquely?

As you can see, the list of considerations gets long quickly and can get quite complicated. Again, we suggest that the serious student look to Edwards and Holland (2013) for a full treatment of the process of interviewing.

Another data collection technique involves group interviews that would include focus groups. Group interviews are a different animal altogether from individual interviews. In addition to the considerations about the scope and mechanics of the process, you must now deal with the dynamics of the group. Will you set the interviewer role as one of asking questions, moderating the discussion, facilitating the discussion, or all three? How will you control the dynamics of the interactions between group members, particularly if strong differences of opinion or conflict arise? Dealing with group dynamics requires a highly skilled person who is trained in such processes.

Another technique is observation. Observations can be on a continuum from *obtrusive* to *unobtrusive* and variations between. In an obtrusive observation, the observer is known to the individuals being observed and is, in some way, part of the context where the behaviors being observed are displayed, but is not influencing the behaviors. This is called **participant observation** (Fetterman, 1998).

An example might help clarify this. In one version of an older observation method called the **chronolog** (Simeonsson & Boyles, 2001), the observer wears a steno mask with a recording drum on the belt (we said it was old!). The observer joins the session, observes the group or family, and talks into the steno mask describing what he or she is seeing; the observations are transcribed later.

While the presence of the observer initially changes the behavior of those observed, eventually, so the thinking goes, the observed will **habituate**—get accustomed—to the presence of the observer and revert to normal behavior. This assumption was the basis for Jane Goodall's observations of chimpanzees and Dian Fossey's observations of mountain gorillas. For an entertaining read about naturalistic observation, you might want to read Farley Mowat's *Never Cry Wolf.*

Indirect observations, sometimes called *unobtrusive* or *nonreactive* observations (Webb, Campbell, Schwartz, & Sechrest, 1965), refer to situations where the individual or individuals being observed do not know that they are being observed. Following schoolchildren around a museum to see what exhibits are most interesting to them, or observing problem-solving behavior of adults through a one-way mirror, would be examples of this.

As should be obvious, indirect observations can raise some ethical issues of concern, since subjects are unaware they are being observed because seeking informed consent negates observation of natural behavior. Common sense suggests that the behaviors being observed must be public and available for anybody to observe, and not illegal or dangerous or compromising to the ones being observed. There also should be no attempt to influence the observed behaviors overtly or covertly.

As with any concerns about ethics of research, it would be wise to consult the Federal Regulations, the American Psychological Association (APA) ethical standards, the institutional review board (IRB), and colleagues as sounding boards for ethical guidance. More on this in Chapter 9.

Figure 6.1 gives several more common data collection methods. The mechanics of these should be obvious. **Surveys** and **questionnaires** are common methods of collecting data and can be either quantitative or qualitative. A good survey or questionnaire requires the same complex considerations as suggested under the interviewing description. Many a study has been undone because the survey or questionnaire was not constructed well and thus was misinterpreted by subjects, spoiling the results.

Case studies are another data collection technique. As the title implies, they are studies of cases—simple enough. For examples, Thigpen and Cleckley (1954) studied a multiple-personality case that became the basis for the movie "The Three Faces of Eve," or Erik Erikson's studies of Luther and Gandhi.

Primary and secondary document analysis is common to fields such as history and English as well as other disciplines. It involves the study of existing documents. Primary documents relate directly to the event under study, while secondary documents provide ancillary perspectives to the event. An excellent article by R. A. R. Fraser (1995) deals with Lincoln's death and involves the careful analysis of medical notes taken by the two physicians in attendance (primary documents) after the assassination attempt.

Other notes made by observers, not related to the medical questions Fraser was asking, would be considered secondary, but important, documents. These would give a broader view of the feelings, perspectives, etc., of those who kept vigil at Lincoln's beside through that long night of April 14–15, 1865.

Data Sources

Data sources can be varied. To return to the Fraser study just mentioned, some of his data sources would have been **archival**. Much of the material for Erikson's analyses of Luther and Gandhi would have been archival. The U.S. government, through its archives, libraries, historical repositories, etc., is an excellent source of archival material, as are governments around the world.

Archival data sources always offer the opportunity of discovering new material that markedly changes perceptions of an event. For example, Preston (2015) uncovered never-before-seen material hiding deep in the archives of depositories in England that markedly changed the historical account of Braddock's defeat at the hands of the French and Indians in 1755 at the Battle of the Monongahela.

To repeat, as with most all archival material, you are the mercy of the accuracy of the individual or individuals who first recorded the data.

Naturalistic settings are another good source for qualitative data. Again, this brings to mind the studies by Jane Goodall and Dian Fossey. Margaret Mead's studies, though ethnographic, were based on **naturalistic observations**. We have mentioned Farley Mowat's study of wolves as an excellent and humorous example of naturalistic observation. May we add the Owens's (1992) excellent study of life on the Kalahari Desert (*Cry of the Kalahari*)?

Existing documents are another commonly used source of data, of particular value in the disciplines of English, religion, and history. Such documents can provide a rich source of qualitative data. We will return to existing documents as a source of knowledge when we get into the next section, with a particular focus on the Bible.

Another data source might be secondary sources. By clarification, primary might include eyewitnesses or people involved in the event, while secondary sources represent data collected by another individual or entity after the event or someone not directly involved in the event. This can get somewhat confusing—we must keep temporal sequences in mind.

For example, an individual in the Twin Towers on 9/11 is a primary source for information about the actual collisions of the airplanes and the aftermath. A firefighter's recount of that event is secondary to the actual collision, but primary to the rescue efforts. A physician treating injured people at a local hospital would be a secondary source for the rescue attempts, but a primary source for the types of injuries caused by the attack.

Examples of secondary sources might include medical records, legal records, minutes from civic or governmental groups, other qualitative research reports, etc. Field notes from a previous researcher would constitute a secondary source in comparison to your own current field notes. A researcher could match his or her own findings and observations with those of the previous individual, based on a comparison of those field notes.

Primary and secondary sources need not be written sources. Photographs, videos, news recordings, taped recordings, etc., could all be used by the qualitative researcher. It is probably fair to say that the data sources for qualitative methodologies are only as limited as the imagination of the researcher.

Apart from the general categories of archival, naturalistic, primary, secondary, and document-based, any of the common data collection methods presented in Figure 6.1 would qualify as data sources. That means you can add case studies, interviews, surveys, questionnaires, direct and indirect observations, and focus groups to the list.

Data Handling

In Figure 6.1, we present basic generic data-handling methods. Data can be coded (key words or phrases coded, etc.), categorized (lumping common statements or themes together), labeled (basically tagging or condensing content), or recombined. Within these generic data handling methods, there are many variations.

Obviously, organizing and processing qualitative data is not as straightforward as plugging quantitative data into a computer and generating output. The ways in which data can be handled and analyzed are limited only by the imagination of the researcher. Currently, there are about 50 identified qualitative analysis techniques (see Section 6.4; Mayer, 2015; Tesch, 1990) with more techniques being developed as computer support grows more flexible and sophisticated.

It is well beyond the scope of this chapter to cover all 50 techniques. What we can do is provide a general indication of the stages in the process of getting qualitative data to the point of analysis.

We can start by dividing the process into three stages. The first stage, obviously, would involve **recording** the data initially, **transcribing** the data later if needed, and then **reducing** the data as much as possible to a manageable amount. Since qualitative data can be extensive, from interviews and other sources, some means of initial reduction will probably be necessary. A primary way of doing this is to reduce or eliminate material not directly related to the research question(s) or emerging themes.

Stage two involves making an initial analysis of the data. This could involve looking for themes or patterns, deviations within the data, stories that might emerge from the data, or date-time sequences that appear to be important. This is essentially a general review of the data and an attempt to identify patterns in the data.

Once past the general review above, one might begin to look for more-specific, targeted, focused, informative themes, patterns, etc. This is where those 50 techniques come into play. We can, however, divide this analysis process into six major components.

The first is called **content analysis**. Content analysis is driven by the research question and the expectations of the researcher. The researcher examines the data for what he or she believes to be in the data, looking for trends, patterns, themes, etc., that should exist. In other words, he or she has predetermined ideas about what should be there, based on previous research, experience, etc.

Data can also be analyzed without predetermined expectations; in essence, letting the data "reveal" themselves and any existing trends, patterns, themes, or theories. This is called **grounded theory analysis** because any findings are "grounded" in the data and reveal themselves as the data are reviewed and analyzed.

In grounded theory analysis, the researcher is interested in the data forming the theory—in the data forming the patterns. An example of this might relate to veterans returning from the Middle East theaters. There are a plethora of services for veterans, ranging from educational support to medical and psychiatric support. Even with all these supports, though, the suicide rate is alarmingly high among this group. One might ask "why,'" given all the available services. One might then ask, how/can/do the veterans find those services?

A grounded theory approach would collect data and let the data be the source of the how/can/do of the issue. The data would not be examined for assumed patterns of how veterans hook up with services—starting with those assumed patterns might block awareness of alternate patterns that may be more descriptive of the actual process of reintegration into services. Rather than the researcher being a sleuth with an idea (content analysis), the researcher is a sleuth *looking for* a possible idea (grounded theory analysis).

Triangulation is an approach that applies to both quantitative and qualitative data analysis (Graue, 2015; Mayer, 2015). Triangulation simply means bringing more than one source of data to bear on the research question. The more overlapping sources for the research question, particularly if they are mutually supportive, the more strength one can apply to the findings based on that overlap.

As an example, suppose we have photographic evidence that aligns with interview data that supports any findings drawn from the interview data that also confirms observational data. This is triangulation, where other data sources are confirming one another. The more data sources triangulating on each other, the stronger the findings.

Computer programs are becoming much more useful at all stages of handling qualitative data. It is well beyond the scope of this book to go into such programs, so we urge readers to search the Internet for qualitative analysis programs. Apart from the software of specific programs, computer programs generally will aid in storing material, searching material and retrieving information, coding, theoretical modeling, and—of course—writing reports.

Think of the library databases as an analogy to coding, identifying, and retrieving material. Put in search words, and the computer finds related articles and sources. Computer programs to date, however, cannot replace the human brain. They can narrow the data, highlight certain data, help condense by coding, but whether these functions are valuable is determined by the researcher's gray material.

Ethnographic methods can involve participant observations, indirect observation, interviews, focus groups, etc. It is a broad field, intended for studying a culture in all of its manifestations. Any of the data collection methods can find a home in ethnographic investigations. **Phenomenology**, the study of subjective experiences, can be part of an ethnographic investigation. Phenomenology may focus more on the underlying maintenance structures for those experiences, such as mores, philosophy of a people, etc. Did you see *The Last of the Mohicans* with Daniel Day Lewis (or read James Fenimore Cooper's original book)? An opening scene shows Natty, Uncas, and Chingachgook hunting. When the deer is killed, they offer a prayer to the Great Spirit for the gift of the deer and to the spirit of the deer—an underlying philosophy of life for the Mohicans. That is phenomenology!

An interesting area of data analysis is **hermeneutics**. This technique has been applied extensively to the Bible, as mentioned earlier. The intent is to identify changes in organization, expression, content, identity, voice, cultural underpinnings, verb use, idioms, references, etc., in the textual language that might suggest a change in authors or multiple authors for one book, perhaps spread over time. As a result of such inquiry, scholars have suggested that multiple hands were involved in the construction of the narrative texts of the Bible, such as the Priestly tradition versus the Old Epic tradition, Yahwist narratives versus Elohist narratives, etc., in the Old Testament (Anderson, 1986).

Such studies suggest Mark was written first, and that Mark, Matthew, and Luke shared both overlapping material and distinctive nonoverlapping materials predating the actual writings that are referred to as Q, M, and L sources (Johnson, 1999). One of your authors completed a master's in religion a few years back and found the erudition, training, language skills, cultural knowledge, and general knowledge of those applying hermeneutics to the Bible to be quite an incredible exercise in scholarship.

Generalization

With qualitative data, we are not attempting to generalize from a sample back to a population. However, there is the hope that qualitative data will generalize into larger patterns, trends, or theories that then can be examined with alternate groups, events, or phenomena, or subjected to quantitative analysis. To the extent that there is a sense of broadening the applicability of qualitative findings, there is the notion of "generalization."

Limitations of Qualitative Methods

Certainly, the dyed-in-the-wool quantitative researchers might suggest that qualitative methods are a bit too loosey-goosey for their tastes. The openness of qualitative research also can leave some researchers feeling adrift and uncertain about goals, direction, veracity of findings, etc. Qualitative research depends much more heavily on researcher expertise and subjectivity. Looking for patterns and paths in qualitative data is much like a clinical psychologist looking for meaning in drawings by patients. Another limitation is that you might have to conduct many, many studies before a meaningful theory begins to emerge.

While these are all just limitations, they tend to imply an either-or perspective relative to quantitative and qualitative research. That is, you can do either quantitative, with its limitations, or qualitative, with its limitations. The picture may become brighter if, as suggested earlier, these two methods are looked at as complementary.

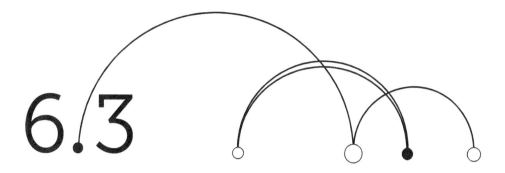

6.3

Qualitative methods: Timeframes

Naturalistic observations, in particular, in remote and wild areas can cover spans of years and even decades. Jane Goodall started her research on chimpanzees in 1960. Dian Fossey's research with the mountain gorillas lasted 18 years before her untimely death. When looking at the research of these two pioneers, we can note that it is both longitudinal and cross-sectional—longitudinal in that they followed their respective subjects for many years; cross-sectional in that they observed development simultaneously across all the possible different age, gender, and developmental stages. Margaret Mead's cultural anthropology research would meet the same general time guidelines (Mead, 1972).

Timeframes in qualitative research are not fixed—they are at the discretion of the researcher. By the very nature of the methods employed, one can anticipate that they will be quite varied across studies.

.

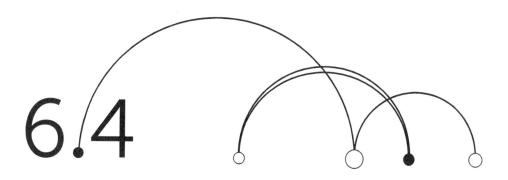

6.4

Broad types of qualitative methods

As mentioned, there are some 50 qualitative analysis methods. The list is too long to include as a whole in the confines of this chapter. We will, however, give the reader a segment of that list for personal edification (Table 6.2; Tesch, 1990; Mayer, 2015, p. 57).

TABLE 6.2 Sample List of Some of the Qualitative Analysis Methods

Action research	Document study	Ethnomethodology	Imaginal psychology
Case study	Ecological psychology	Life history study	Intense evaluating
Clinical research	Educational	Naturalistic inquiry	Participant observation
Cognitive anthropology	Connoisseurship and criticism	Panel research	Participative research
Collaborative inquiry	Interpretive	Ethnoscience	Phenomenography
Content analysis	Interactionism	Experimental psychology	Phenomenology
Dialogical research	Educational ethnography	Field study	Qualitative evaluation
Conversation analysis	Ethnographic content analysis	Focus group	Structural ethnography
Delphi study	Interpretive human studies	Grounded theory	Symbolic interactionism
Descriptive research	Ethnography	Hermeneutics	Transcendental realism
Direct research	Ethnography of communication	Heuristic research	Transformative research
Discourse analysis	Oral history	Holistic ethnography	Transactionalism

Where to Next?

This completes our review of the two main types of research methods, quantitative and qualitative. Next we will learn how one (or both) of these methods fits into the overall research process.

Key Terms

- Archival data
- Chronolog
- Coding
- Content analysis
- Dialogue
- Ethnography
- Freudian theory
- Grounded theory analysis
- Habituate
- Hermeneutics

- Indirect observation
- Mixed methods
- Naturalistic setting
- Objective knowledge
- Participant observation
- Phenomenology
- Primary document
- Qualitative methods
- Questionnaire
- Reducing data

- Secondary document
- Subjective knowledge
- Survey
- Theme
- Transcribing
- Triangulation
- Trustworthy
- What, how, and why

Questions

1. Why does Freudian theory qualify as a qualitative theory?

2. Describe a direct and indirect observation method related to pilot training.

3. You wish to gather qualitative data on air traffic controllers in training. Generate three potential sources of data that can be used to triangulate your data.

4. What happened to Dian Fossey and why?

5. The rationale for using the chronolog method as an indirect observation process is based on habituation. What does habituation mean in this instance?

6. If you were going to do a qualitative research study, what would your area of interest be and what would your research question be?

7. Design a case study—that is, what is a case study, who/what would you study, and how would you set up the study?

8. Identify a phenomenological variable that you could study.

9. Design a mixed-methods approach to a particular research question. Be specific— what is the question, and which part of the investigation would be qualitative and which part quantitative?

10. What do Dian Fossey, Jane Goodall, and Margaret Mead have in common?

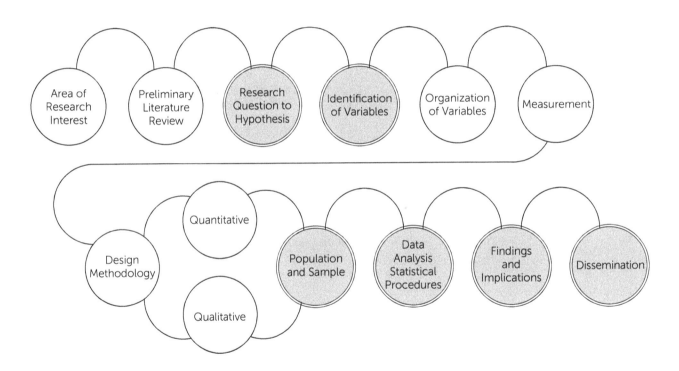

Area of Research Interest

Preliminary Literature Review

Research Question to Hypothesis

Identification of Variables

Organization of Variables

Measurement

Design Methodology

Quantitative

Qualitative

Population and Sample

Data Analysis Statistical Procedures

Findings and Implications

Dissemination

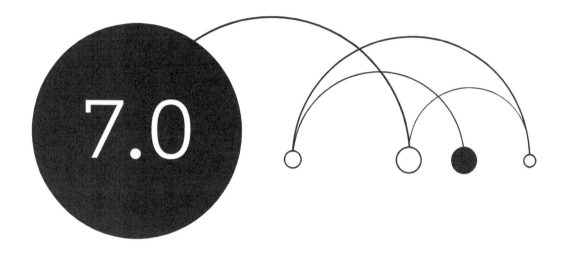

Putting It All Together:
From Process to Product

A lottery is a tax on people who do not understand research.

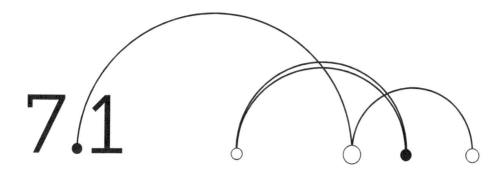

7.1

Arriving at a research question

The research question, as we have said, is what drives the research. Questions can arise from many places—existing theories, constructs, other research, or events around you. In some ways, research questions that arise from what is going on around you may be the most interesting, since those events apparently have piqued your curiosity, stimulated questions, and may affect you directly or indirectly.

Events that can lead to research are not hard to find. A number of years ago, a graduate student of one of your authors found a survey that had been used during Desert Storm to assess public acceptance of the action taken during that conflict. That student, curious, re-used the same survey to assess the acceptability of the Iraq invasion during the Bush administration. The findings were quite interesting and, because complete data were available from Desert Storm, the two studies were comparable. Both the student and his thesis committee became very excited about the research, its applicability to current events, and the rather creative nature of the study.

What follows is a case in point. This study arose out of a current event. The event both stimulated our curiosity and possessed a very serious and immediate urgency. It is quantitative, but we will point out the qualitative differences a bit later. Here we go!

Identifying an Area of Interest

Two of your authors were sitting at a restaurant high above the Ashley River in Charleston, SC, discussing their research schedule for the next several months in reference to several projects underway. During the course of the conversation, Germanwings Flight 9525, which was still in the news from the crash just a few short days before, came up.

As you may recall, this was a flight from Barcelona, Spain, to the Düsseldorf airport in Germany. During the flight, the captain left the flight deck and, while he was out, the co-pilot locked the door. Shortly after, the plane began a rapid unauthorized descent from its 38,000-foot cruising altitude and crashed, killing all on board.

When flight data recorders were recovered, the captain was heard trying to make contact with the co-pilot through the cockpit door, then trying to break open the door. In the background, passengers were heard screaming. This was a very tragic occurrence and clearly a case of suicide by aircraft. As we talked about this event, we had a number of questions: How could this have happened? How often has it happened? How would you know in advance if a pilot was suicidal? Would it be possible to prevent such an occurrence? These were questions others were asking, as well. The only problem: There were no good answers.

As our discussion of the Germanwings crash progressed, the question arose of whether we had any suicide data in the National Transportation Safety Board (NTSB) crash database we had been using to study gender differences in male and female pilots. We examined the data and, indeed, found suicide as one of the coded items for cause-of-crash. This was exciting!

Given the nature of the data in the particular database we were using, research questions would be limited to pilot characteristics such as gender, age, hours of flying experience, certifications, conditions at the time of the crash, etc. We did not have any medical or psychological data to work with. However, within the framework of what we did have, we wondered whether we could provide some description of the characteristics of pilots who commit suicide by aircraft.

Out of this arose two research questions: 1) What are the characteristics of pilots who commit suicide by aircraft and 2) Is it possible to differentiate suicidal from nonsuicidal pilots based on those characteristics?

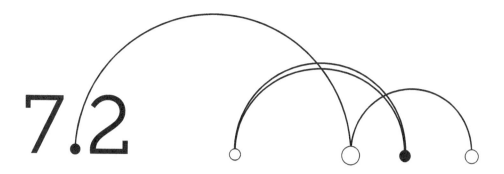

7.2

Validation of the research question, or questions: Preliminary literature review

By the time we finished our discussions that evening, we had two good research questions in hand. They were focused, simple, parsimonious, operationalizable, and, we hoped, worth pursuing. Worth pursuing, you say? What do we mean by that?

We had several preliminary tasks ahead of us before we went anywhere with the research. The very first task was to find out what was already known about aircraft-assisted suicide. Had such studies been conducted? We needed to delve into the literature to see what we would find. For all we knew, somebody had already conducted the study using the same database we had, or some studies had been conducted but were incomplete in terms of which variables they were able to focus on. We simply did not know, and we had to know—you do not want to duplicate, unknowingly, something that has already been done.

Using library resources from our respective institutions, we were able to isolate all research articles on aircraft-assisted suicide. There were none that we could find! Notice we qualified this with "that we could find." There are literally tons of research article databases out there in cyberspace. It is unlikely that any one—or two—institutions will have all of them covered. In that preliminary search, you want to make a good faith and diligent effort to isolate any similar studies. That is a minimum expectation.

The next question was what variables were important to examine with regard to aircraft-assisted suicide. Since there were no studies specific to aircraft-assisted suicide, we were on our own—but not completely. There is a very rich literature on suicide in general, suicide by automobile, gender differences in suicide, etc. This would be the basis for our selection of variables and would guide our efforts.

Without the foundation of **previous research**, you cannot tell where your study will fit into the flow of findings. Even if that research is from allied but closely related fields, it gives you a base for starting your investigation. As stated in Chapter 5, research often progresses by small steps, expanding the boundaries of our knowledge. You do not want to spend time on a study that ends up simply duplicating, unknowingly, results found by other studies. That is working backward, not forward.

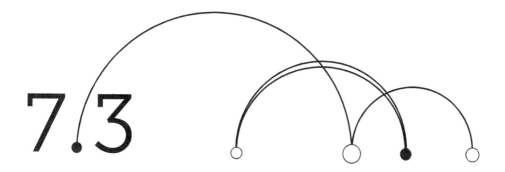

7.3

From variable selection and preliminary literature review to identification of potential sample

In our case, we already had the NTSB crash database. We did not have to identify potential subjects or measurements—an advantage of **archival databases**. We did have to identify those variables in the database that were germane to the research questions, since not all variables we had appeared to be relevant, such as fixed or collapsible landing gear. We also had **narrative summaries** from the aircraft accident investigators and needed to read those summaries for more specific-information on all suicide-related crashes. There was valuable information in the narratives, such as whether the pilot was on prescription medication, toxicology reports following the crash, etc., all of which shows up in research on suicide by car and similar topics.

While archival databases are quite convenient, they do have limitations. It was not possible to go back and gather additional information that we thought might be important. We also were dependent on the aircraft accident investigators regarding the thoroughness and attention paid to the cause of the crash, and on whoever actually coded the data into the system having done so accurately. Both factors could have introduced errors that we would know nothing about.

When archival databases are not available, one is tasked with gathering new data. This is an exercise in creativity and good sleuthing. Again, you must identify variables you will be interested in based on your preliminary literature review. You will have to identify some means of operationalizing those variables for quantitative research. The quantification may well involve tracking down reliable and valid measures.

Then there is the issue of finding subjects pertinent to the research questions. And, of course, you have to run all of this through an **Institutional Review Board (IRB)** committee for review and approval *before* collecting any data.

As an example of a nonarchived data set, a number of years ago, one of your authors was involved in a large study looking at depression in children that spanned a number of institutions in several states. At that time, the assumptions were that depression in children and adolescents was just like adult depression and, therefore, the criteria for adults in the Diagnostic and Statistical Manual of Mental Disorders would simply extend downward. Based on the experience of a number of us working in psychiatric settings, though, we did not think that children and adolescents manifested depression in the same manner as adults.

We set out to investigate this and, of course, initial considerations revolved around existing studies—none—and, in light of the absence of relevant prior studies with children and adolescents, what could we glean from the myriad of studies with adults and the theoretical formulations around depression? Secondary questions revolved around possible measurements we might use that possibly could differentiate any differences, if there were differences, between children/adolescents and adults, to include quantification of diagnostic criteria, intake interview notes, etc.

The hunt was on. Additional considerations included possible differences by developmental ages, specifically between children and adolescents, by gender, etc. We did have some theoretical direction from researchers, such as **John Bowlby**'s studies on attachment and separation of children (Bowlby, 1980) and, of course, an abundance of adult literature.

That literature—both theoretical and applied—served as our starting point and an indicator of directions that might be fruitful. It took more than two years of exploratory qualitative and quantitative research before we began to identify patterns and trends that were worth following. Simply identifying the layout of the land was a long process.

We would like to step aside for a minute and point something out. If you are a Sherlock Holmes fan, you may be a prime candidate for a career in research. Our use of the word "sleuthing" was no accident—research can be very much like sleuthing. It is challenging intellectually, results in some great camaraderie with colleagues involved in the same or similar research, and yields a profound sense of triumph when things begin to fall into place and make sense. As strange as this may sound, it is a thrilling enterprise. Again, think of Sherlock Holmes: "Watson, the hunt is on!"

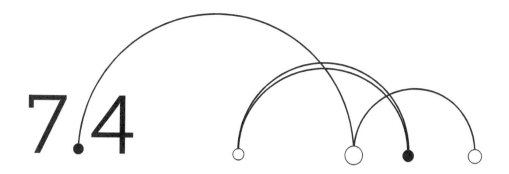

7.4

From sample identification and variables to formal literature review

It is now time to take that preliminary literature review and turn it into a formal review. This can be done while the study is going through an IRB review and while you are tracking down measurements. It is good to get started on this early, since it usually will be the longest part of any research paper, particularly if that research paper is a thesis, dissertation, or capstone project. Your literature review is also critical to your discussion of your results.

We will not say much more here, since this has been thoroughly covered in Chapter 2. Keep in mind that literature reviews must be organized in a logical manner, should never be rambling presentations of past studies in the order they were found, and should lead from the broad to the specific where the hypothesis drops out of the review naturally. Think of it as winding your way down through a funnel, from the broad top part, ever narrowing, to the final point of dispensing what was placed in the funnel—in the case of literature, the hypothesis.

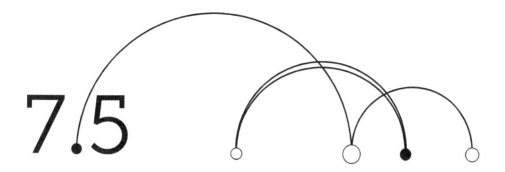

7.5

Onward to data sources and collection techniques

Your IRB has been formally approved. Your formal literature review is progressing nicely. You have operationalized your variables in the case of quantitative research. You have identified your subjects and have the proper mechanisms in place to obtain informed consent (see Section 9.3). You are ready to go.

For our suicide-by-aircraft study, we isolated all cases in the archival database that had been designated by crash investigators as involving suicide. We then tagged these cases numerically (a nominal tagging) to distinguish from nonsuicidal cases: We tagged suicide cases with 1 for suicide and nonsuicide cases with 2 for no suicide. This created the groups for our second research question—the issue of distinguishing one group from the other.

We added additional variables to each of the suicide cases based on an analysis of the narrative summary. This extended the variables list for those cases.

The very first step in quantitative research, once data have been entered into an analysis program, is to run **frequency distributions** to check the **range** and accuracy of your data. We found cases in our very large database (more than 67,000 entries from 1984 to 2014) where a subject held a pilot's license but had zero hours of flying experience. Obviously, this was an error (missing data, in this case) and relates back to the issue of the accuracy of any archival database.

Frequency distributions aid greatly in identifying such errors. Remember: Whether your data are quantitative or qualitative, your research study will only be as good as the data.

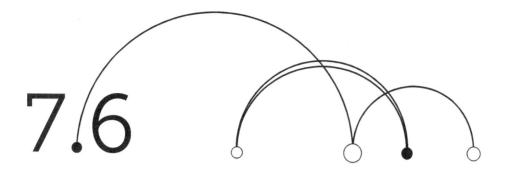

7.6

From data sources to analysis

The research question or questions will drive the statistical analysis. For our study, we had two such questions. The first question was addressed simply through the frequency distributions. We looked at average age and age range of suicidal pilots in the General Aviation (GA) data set (there were no suicide-identified crashes in the database for commercial air operations). We examined flying hours, gender, types of licenses held, etc. When we finished, we had a profile.

The next step was to see if the profile of suicidal pilots differentiated them from nonsuicidal pilots. We ran a **discriminant function analysis** (this is a multivariate technique). We created one discriminant function equation that identified suicidal pilots at 98% accuracy in the database. Wow!

The only problem: There was also a 95% suicidal identification rate for nonsuicidal pilots. This is the false positive-false negative conundrum. We could certainly identify, post-accident, those pilots who committed aircraft-assisted suicide. However, we could not differentiate, based on pilot characteristics, those who did and those who did not.

This study was published because it was a "first" as far as we or the journal reviewers and editor knew. It was a study that might stimulate further studies of a more-precise nature and provide the foundation for extending from GA to commercial operations.

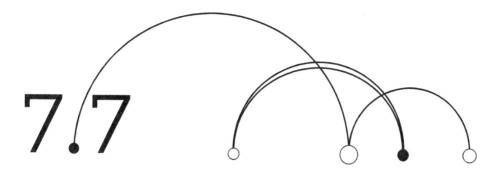

7.7

Quantitative and qualitative differences

How would a qualitative study differ from the quantitative one described? The primary differences would be with Section 7.1, 7.5, and 7.6. All else would be essentially the same.

For Section 7.1, your research questions would be qualitative in nature—the **how, what, why**. They would be more subjective, perhaps aiming to get some idea of the pilot's state of mind before the suicidal crash occurred. Instead of using an archival database as we did, you might generate a **questionnaire** that you could use as a guide to interviewing crash site investigators, next of kin, etc., to establish the state of mind and the medical and psychological condition of the pilot, recent stressors, etc. (If you go back and look at Edwin Schneidman's (2004) psychological autopsy of suicides, you will find a wonderful qualitative and quantitative model that could be modified to use in such a study.)

For Section 7.5, you would collect data per your questionnaire. This might involve, as suggested, **interviews**. It might also involve looking at archival data available through the NTSB database, but with more of a qualitative eye to the data than a quantitative one.

For Section 7.6, you would not use statistical procedures. Rather, you might look at data condensation or arrangement of data around what appear to be common themes (**content analysis**), or you might let the data speak for itself and see what it presents to you (**grounded theory analysis**). You might use both of these, or any of the other applicable analysis techniques mentioned in Chapter 6.

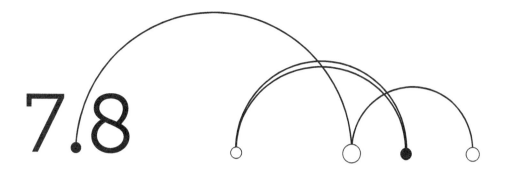

7.8

From write-up to dissemination

Y ou have now learned everything (well, maybe not *everything*) you need to know about how to conduct research, and learned that it is a process. The final step in the process is to distribute your findings.

If you are a student, your main avenue for distributing your findings is submitting your paper to your professor for a grade. However, sometimes your research, if conducted properly and even if *just done for a class*, may hold value for the scientific community. This section will provide an overview of the publishing process and how to prepare your research for publication and presentation, should you choose to go those routes.

This book has been written for the beginner researcher and we have covered many topics. The purpose of this section is not to teach you *how to research* (we have already done that), but to give you ideas about how to get your research published when it is ready.

Presenting at Conferences

Where most students typically start to publicize their work is at a conference. Conferences are great because you can meet other researchers in your field and get immediate feedback from experts. The best part: Conferences are normally held in some cool places. Some of your authors have presented in places like St. Petersburg, Russia; Venice, Italy; Vienna, Austria; Amsterdam and Groningen, The Netherlands; Berlin, Germany; Boston, MA; Washington, DC; Portland, OR; San Jose and Long Beach, CA; and even New Orleans during the NCAA finals and Mardi Gras. How cool is that? Have research, will travel!

As mentioned above, attending conferences provides an avenue to receiving feedback on your research so you can improve your methodology or plug holes that may cause a manuscript to be rejected by a journal. True story: While one of the authors was conducting his dissertation research, he presented some of the preliminary data at a conference. During the presentation, one of the people in the back of the room kept asking questions—and they were good ones. After the presentation, your author approached the questioner to talk about his comments and try to learn more—that's part of the research process.

When your author introduced himself and heard the questioner's name, he immediately realized that he had cited this guy throughout the literature review and had read many of this colleague's papers—he truly was an expert on the author's dissertation topic. They had a long conversation about the research and he gave your author some good ideas and additional papers to read. His guidance was invaluable to your author's dissertation, and they are still in contact today.

At most conferences, there will be both **presentations** and **poster sessions**. We would recommend that when you are ready to present your research for the first time, you go for the poster session. This gives you some exposure, and is a lot less intimidating than presenting to a room full of people. There also usually are more slots for posters than for presentations, so your chance of acceptance is higher.

Take the feedback received during the conference and update your research accordingly before submitting to a journal—you will end up with a better product.

Proceedings

Some conferences publish presented papers in a **conference proceedings**. Proceedings are generally not peer-reviewed, other than the review that led to acceptance, and so are a good way to get your research out there for others to use. Nevertheless, you will still need to make sure that you have a good, solid product to submit.

Selecting a Journal

After you have presented at a conference and refined your paper in response to feedback from that experience, it is then time to try to get it published. One of the hard parts is finding the right journal for your manuscript. The best way is to review a number of journals to find a journal that is a good match between its purpose and your line of research and writing style.

Most journals will have a stated purpose and are looking for articles that match that purpose. If the journal is titled *Aerodynamics*, then a paper on ramp safety, no matter how good, is not relevant to that journal and will be rejected. You are just wasting your and the editor's time. Our pilot suicide paper was published in *Suicide and Life-Threatening Behaviors*—very *apropos*, given the topic!

Types of journals

Peer-reviewed vs. **non-peer-reviewed**—that is the question. Academic journals can be broken into four types:, peer-reviewed, peer-/**editor-reviewed**, non-peer-reviewed, and proceedings.

Peer-reviewed journals require that your manuscript undergo a review by several experts in the field (normally blind or double-blind reviews, in which the reviewers do not know who wrote the paper). Reviewers ensure that your paper meets the highest research standards, is well-written, includes a methodology that is generally accepted in the field, and provides results that are supported by previous research and findings on the subject.

On the other hand, there are the non-peer- or editor-reviewed journals. While a paper that undergoes only an editor's review is not seen as meeting the same quality standards of a peer-reviewed article, editors of legitimate journals normally do not simply publish any and every manuscript they receive. An editor wants to make sure that papers or articles that are published are solid, well-researched, and well-written.

Almost all of the non-peer-reviewed journals require a fee from you to publish your work and do not use a review process. Since your findings do not go through any review, they may be considered dubious by the academic community.

A word about fees. Many respected and well-regarded journals charge a per-page publishing fee. Let's face it: Academic journals do not make up a bulk of what you will find at Barnes and Noble or on Amazon. The primary markets are academic institution libraries and, perhaps, professionals in the field who subscribe as a means of keeping up with what is going on. Given such low volume, costs must be offset. Just because you are asked to pay does not mean that the journal equates to a vanity press type of publication. Just be cautious and do some research about a publication before committing funds to getting published.

Proceedings are simply collections of papers presented at conferences: By presenting, you usually are included in the event proceedings. You may be asked to give permission for your manuscript to be used, but you do not have to go through a formal submission—or revision—process.

Preparing the Manuscript

In Chapter 2, we provided an overview of how to format your paper per APA style. As noted, there are many types of formatting systems, so review the submission requirements—usually published on the journal website—and follow them closely to enhance your changes of publication. Most journals these days will accept **electronic submissions**, but some still want a printed copy of your manuscript as well.

Make sure that your paper outlines the research process you followed and that the results support your research question. The research question should be based on previous research.

When you are ready, submit your manuscript and then be prepared to wait. A few journals will review your manuscript within a month or two, but this is not the norm. Expect four to six months. Do not get frustrated and submit to another journal—that is unethical! You have to formally withdraw your manuscript or wait for a response from that journal before submitting elsewhere.

Notification About Your Manuscript

After review of your manuscript, journals will normally either reject it outright, accept it with revisions, or accept without revisions.

Outright rejection

Rejection of your manuscript happens. The higher the quality of the journal, the more likely your manuscript is to be rejected. You can always expect rejection if your research does not match the type of articles the journal is interested in. The good thing is that most journals will tell you why they rejected your study, so you can update your manuscript/research based on reviewer comments and try again. Updating will strengthen your paper!

There could be a number of reasons for rejection of your manuscript.

- Manuscript/research does not match the purpose of journal. The way to correct this is to submit to a more-relevant journal.
- Research design, methodology, analysis, or interpretation is flawed or incomplete.
- Manuscript not clearly written or argument not clearly laid out or backed by facts, and/or explanation of your findings does not relate/support your research question.
- Improper format and/or citations.
- Grammatically incorrect sentences and spelling errors.
- Literature review does not reflect the breadth and depth of the topic.

Acceptance with revisions

Journal reviewers thought the article fit within the purpose of the journal and that your research methodology was sound—or mostly sound. You may be asked to expand on a section, or conduct more analysis to plug a hole (these days, most requests are about the effect sizes), but overall your manuscript is good, so you can update and resubmit. The editor cannot guarantee acceptance of a revised manuscript, but if you make the suggested corrections in a timely manner, it increases your chances of getting accepted. You may want to consider submitting a change matrix, listing the reviewer's comments in one column and how you addressed the comments in the other column.

If your manuscript comes back again, make the new suggested corrections and return the revised version as soon as possible. Don't get frustrated and go to another journal; you'll just start the process over from the beginning.

Once your manuscript gets accepted, congratulations: You're a published author. Now you can put it on your résumé!

Where to Next?

Next, we will look at some limitations and how to avoid defective research. We will explore such exciting subjects as reliability and validity, and the various types of limitations that may affect your research. But first, a quick review of some key terms covered in this chapter.

Key Terms

- Acceptance
- Aircraft-assisted suicide
- Analysis
- Archival data
- Area of interest
- Collection techniques
- Conferences
- Content analysis
- Data set
- Data sources
- Discriminant function analysis
- Editor review
- Electronic submissions
- Frequency distribution
- Grounded theory analysis
- Institutional Review Board (IRB)
- Interviews
- Journals
- Literature review
- Narrative summaries
- Peer review
- Poster sessions
- Presentations
- Previous research
- Proceedings
- Questionnaire
- Range
- Rejection
- Sample identification
- Submissions

Questions

1. What is the driving force behind your study and what aspects of the study are determined by that driving force?

2. Outline, briefly, the sequential steps in conducting a study.

3. How does a preliminary literature review aid your study?

4. What does a preliminary literature review help you do?

5. Why is a poster session a good place to start when you are ready to present your research?

6. Name at least two reasons why it is valuable for researchers to attend conferences.

7. Distinguish between the different levels of review for article submissions to journals.

8. Find and list some other guidelines for manuscript preparation besides APA.

9. Pick three journals in your area of study and find the acceptance/rejection rate for each.

10. Identify two journals that align with the content area of your research.

11. What are the most likely reasons for rejection of your research for presentation or publication?

12. How do proceedings publications differ from journal publications?

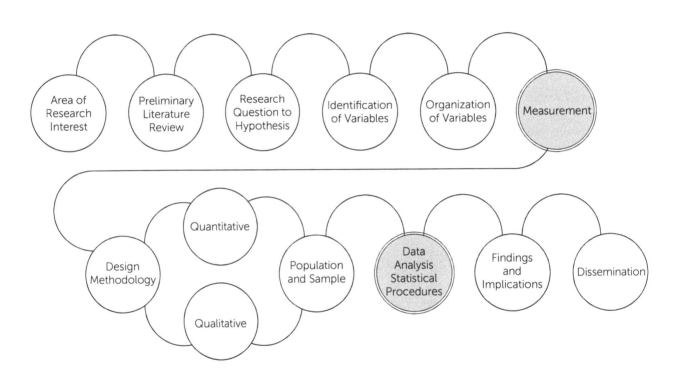

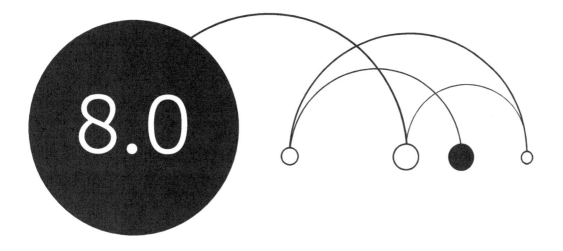

Limitations: How to Avoid Defective Research

A researcher is someone who insists on being certain about uncertainty, can draw a precise line from an unwarranted assumption to a foregone conclusion, and whose lifetime ambition is to be wrong 5% of the time.

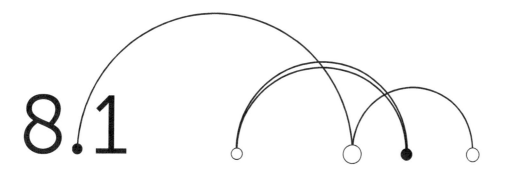

8.1

Let's get started

A number of years ago, a graduate student was doing her thesis on a subtest of the Wechsler Intelligence Scale for Children (WISC), focusing specifically on the revision made to a subtest from the third edition of the WISC to the fourth edition. It was a very clever study that targeted the basic psychometrics of the WISC; in fact, it would have been a rather profound study. Basically, it involved administering that particular subtest under three conditions: according to the format from the third edition, the format from the fourth edition, and a revised format that she felt would be more appropriate, given that items on the subtest were supposed to be arranged hierarchically by level of difficulty. She used sixth-grade students, randomly assigned to one of the three format sequences.

This was a basic experimental design, very straightforward, with no quasi-experimental variables. She did find that the new arrangement of items received a significantly higher score than the other two, thus suggesting that the item sequence on the new WISC might not be arranged according to a hierarchy of difficulty. Significant findings, and noteworthy as well!

Of course, **extraneous variables** potentially related to IQ must be controlled, such as gender (this was a right-parietal task), age (think Piaget's stages of development), and IQ (obviously, since this was an IQ test). Since she could not administer an IQ test to all 60 subjects (20 **randomly assigned** to each sequence), she used grade point average (GPA) as an IQ-related variable.

To make a long story short, there was a significant difference in GPA, with the group receiving the new sequence the highest of the three. Since that particular group was significantly higher on GPA, did the sequence lead to higher scores or were the higher scores related to the significantly higher GPA of that group—and thus potentially higher IQ? There was no way to know. A neat study undone.

Her thesis was well conducted and we did not let this defeat what she had done, so she did defend and graduate. She was undone by sampling error. The point of this story is that many things out there, if not attended to, can sneak up on you when you are least aware and send your study down the tubes. Each of the following sections will present limitations or errors, or both, that can undo your study.

As an aside, the quickest remedy for the study described above would have been to randomly assign more subjects to each condition in the hopes that GPA would even out across the groups. Unfortunately, we had exhausted all of our parent permissions (necessary for dependent subjects for any research participation), so we could not increase the number of subjects.

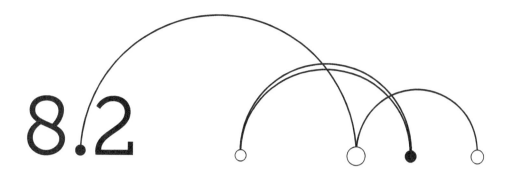

8.2

Revisiting reliability and validity

I t is worth starting back with our old friends, **reliability** and **validity**. Reliability, remember, addresses consistency in measurement and validity addresses measuring what you think you are measuring.

Several years ago, several faculty in the Psychology Department at the Citadel ran a study after the 9/11 strike on the Twin Towers to see how this event affected children. It used a number of well-documented stress, anxiety, and depression scales to assess functioning and a new scale, the Charleston Scale of Self-Perceived Danger (CSSD), was also administered. The CSSD was a straightforward measure assessing the extent to which children perceived that they personally were in danger given the 9/11 attack. The new scale functioned well in explaining part of the variance in the children measured, but the study was rejected for publication because the CSSD had very limited—almost nonexistent—reliability data and, therefore, did not have demonstrated validity (remember, you must demonstrate reliability before you can make any assumptions about validity).

Many editors and reviewers for journals do not like new scales that do not have established reliability and, thus, the potential for validity. The reason for this, which we hope is obvious, is that reliability does not automatically ensure validity (Nunnally, 1967). There are examples of reliable measures that do not have much validity in research—or should we say relevance?

As something fun to do, look up Senator William Proxmire's Golden Fleece Awards. The senator gave annual awards for what he considered the most ridiculous waste of federal research grant money. Some are quite amusing, such as the study of 432 airline stewardess to assess length of buttock—a measure that may have high reliability, but little validity for anything. Another was the search for extraterrestrial life, which could also have high reliability (identification of nonrandom noise from the universe as compared to random noise), but, again, little validity as it relates to anything.

For the 9/11 study, the researchers would have had to conduct many studies with the CSSD looking at convergence with similar instruments, consistency of measurement over time and across groups, etc., to establish the psychometric properties of the CSSD—namely, reliability. That would take time and move the measurement farther from 9/11 in time. The more time that passed, the less sensitivity the instrument would have to the event. Time was our enemy!

Even when an instrument like the CSSD makes intuitive and logical sense (even when it has face validity and apparent content validity), if it does not have demonstrated reliability, it has nothing. Reliability is at the core of any measurement. Without reliability, that measurement is not worth the paper it is printed on or the time it takes to make the measurement, as far as many researchers are concerned.

Enough about reliability and validity. We promise not to mention those two terms again—at least not more than a half-dozen times!

As we look at limitations/errors that can undo your study, we will divide them into categories based on the source. The sources we will look at are **researcher**, **participants**, **environment**, **construct**, **measurement**, **error** of an expected or random nature, and **statistical** constraints. We will then look at ways to guard yourself—and your valuable research—against these problems.

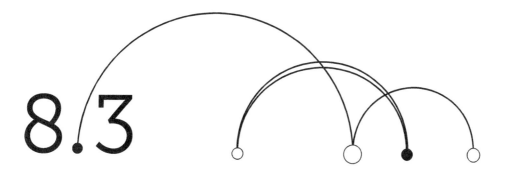

8.3

Researcher constraints

We mentioned Samuel G. Morgan earlier in Chapter 1—the one looking at craniometrics to judge intelligence—remember? His work represents researcher bias on steroids. If you think about it, however, every researcher who states a research question is biased in that s/he holds expectations about how that research will turn out. If we may quote ourselves, from Section 2.2:

As you probably already know, what we typically set out to do when we conduct a research study is to show how the data supports our hypotheses. We do this objectively, honestly, and without bias, of course.

The "without bias, of course" element is the ethical, objective, rational, and open-minded way to approach the research questions. We all strive for that unbiased approach. At the same time, we are invested in our research. We hold our research question dear. It is like a newborn child, and we want it to grow and thrive through our positive and significant findings. We do not want negative results to consign our research question to the anonymity of never being acknowledged.

Perhaps we wax somewhat dramatic here.

A philosopher friend of one of your authors can argue—convincingly so—that you cannot divorce yourself from your expectations of what you hope to find from your research. Therefore, you can never be completely neutral and unbiased. However, awareness of the tendency to become myopically over-invested may be sufficient to at least help us keep our heads clear and minds open to the true nature of the pursuit of science.

In addition to expectations about our program of study and particular research questions, researchers can have expectations about subjects and communicate these expectations in ways that the researcher is not aware of. An area that illustrates this is **child testimony**. It was thought that asking leading questions would result in suggestion-induced memories, i.e., false memories (see Brainerd & Reyna, 2012, for a summary of this area of research). Would researchers inadvertently communicate expected behaviors or responses to subjects, thus biasing results? It is certainly within the realm of possibility.

As researchers, we can also carry biases based on the events or phenomena we are researching. We mentioned the study looking at children's reactions to 9/11. Didn't we all have reactions to that event? Were not those reactions intense? visceral? emotional? As we were doing that study, were we neutral researchers, given the baggage we brought with us from having lived through the event?

Let us suppose you go into neuropsychology and you spend your career working in a VA hospital with soldiers wounded in war zones, fighting against foes that you, as a researcher, may find reprehensible. Will your findings concerning those veterans be unbiased?

In both cases, the feelings may be so intense that we cannot really assume an unbiased approach, even though our training as scientists puts us on alert to possible influences. These are intense settings, but even less-emotionally charged situations can exert an influence on aspects of a study. A few years ago, one of your authors was involved in researching a program called WINGS for Kids—an after-school program aimed at increasing emotional and social intelligence (Gardner, 1993). Having spent years working in inpatient child psychiatry, your author thought this program made good sense. Could these less-intense, but positive, feelings have influenced the results we found in support of the program?

On a more-mundane level, there are issues associated with the care the researcher takes in **recording**, **transcribing** (if necessary), **coding**, and **inputting** data into any computer analysis program. There are also issues of researcher competency in using quantitative or qualitative methodologies, interpreting those findings accurately, and presenting the findings objectively. While we would like to think that all published research is good research, the truth is that some chaff gets through the sieve of peer review.

8.4

Participant constraints

D r. Alfred Kinsey of Indiana University, Bloomington, IN, and his colleagues published two books on human sexuality: *Sexual Behavior in the Human Male* (1938) and *Sexual Behavior in the Human Female* (1953). Both became bestsellers, although that is not the point of mentioning Kinsey. The point is that he and his staff conducted interviews face-to-face. This was at a time when sexual behavior was not a public topic and such behavior based on accepted mores (namely, discretion) was expected in most cases.

Imagine you are in Bloomington and have been asked to participate in the study. You arrive at the Kinseys' institute and are ushered into a room and seated in a chair. A young research assistant comes in and starts asking you questions—very intimate questions—about your sexual behavior, proclivities, number of partners, age at first intercourse, etc. How will you answer those questions? There is a tendency to answer in the **socially appropriate direction**, given the mores of the time. In this case, probably to underreport your actual behavior.

Think about being surveyed on prejudiced behavior. How would you answer the question, "Do you consider yourself to be a prejudiced individual?" Again, the pressure, given today's expectations, would be to answer in a socially appropriate direction.

Why do we do this? We like to be liked and, when asked sensitive questions, the tendency is to answer in a direction that will increase the probability of being liked—the socially appropriate direction, regardless of *actual* internal beliefs or attributions. We practice **impression management**, particularly when in new or unfamiliar situations. Practicing impression management may influence short people to add a bit to their height, heavy people to subtract a few pounds, and people with no singing talent to nevertheless convince themselves that they will be the next "American Idol" superstars. Wanting to be liked and doing what we think will make somebody like us—impression management—is just part of the human condition.

Participants are not passive automatons responding to a research event. They bring with them past experiences, idiosyncrasies, personalities, and other characteristics that make them thinking, observing, and engaged participants in your study. Often, they will try to interpret what you are doing, what you are after, or what you want them to do. This anticipation is variously captured as **participant response bias**, **demand characteristics**, or **reactivity**.

Response bias applies to those characteristics of individuals that predispose them to answer in certain ways. For example, Germans appearing in the Nuremberg trials after World War II were thought to have an authoritarian-responsive personality—that is, they were predisposed to respond to authoritarian people or figures and orders from those such people, and thus, accept the extermination of so many individuals. Optimistic people may respond differently to the same questions than pessimistic, extroverted differently than introverted, depressed differently than nondepressed, etc.

Demand characteristics have two components. One is responding to the perceived demands and expectations of the researcher. Examples would include Stanley Milgram's study on obedience and Solomon Asch's study on conformity. We will talk about the environmental component in the next section.

Reactivity applies to changes in behavior based on participant perceptions of being watched and/or modifications in the environment. Take a look at the famous Hawthorne studies of the 1920s. All watched behavior is potentially modified behavior. Environmental changes can potentially change behavior, as in the Watts riots during a very hot part of the summer. The heat was associated with an escalation in "hot" dialogue and "hot" behaviors.

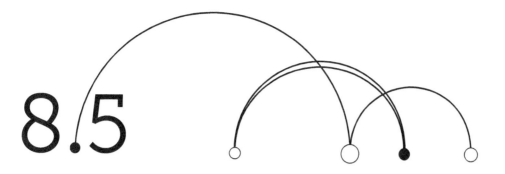

8.5

Environmental constraints

Stanly Milgram thought the prestigious aura of Yale University might be influencing how participants responded in his study—an environmental influence and an environmental demand characteristic. He moved to a nondescript building in New Haven, removed any trappings of Yale, and repeated his studies.

Erving Goffman suggested that we are like Shakespeare's characters, where all the world is a stage. Our behavior changes depending on where we are—at a party, home or work, a sporting event, marriage ceremony, etc. These are all **environmental influences**. Suppose you were to participate in a study where you would be asked questions about how you feel about preservation of the environment. Would you answer differently if asked in the sterile confines of a college laboratory versus at the foot of El Capitan in Yosemite?

Unexpected environmental events also affect measurement. As an example, one of your authors gave a scheduled exam in an undergraduate statistics course. It just so happened that the control board for the chiller in the air conditioning system went on the fritz and the temperature in the classroom was well into the discomfort zone, with humidity to match. Could this have introduced any constraints on grades on the test? Possibly! People tend to not work as well when uncomfortable, particularly on complex tasks (Were the scores then discounted? Absolutely not!).

A loud noise during a memory recall task, an attractive member of the opposite sex walking by during an outdoor performance test, the heat during a marathon—all of these are environmental influences that can affect performance and influence measurement.

Certainly culture, as we will explore in more detail in Section 9.3, represents environmental conditions in a larger and more comprehensive context.

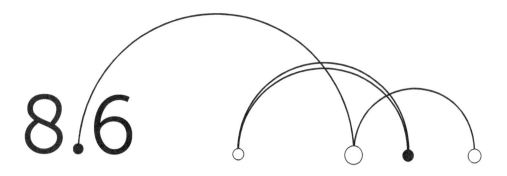

Construct constraints

One of your authors had his astrological chart constructed by the Edgar Cayce Foundation in Virginia Beach, Va. For one year, he followed his chart to see if the predicted convergences of various astrological bodies produced the specified emotional, intellectual, creative, and inspirational fluctuations indicated by the daily chart. The results were equivocal, at best.

There are two primary explanations for this. One: Your author may have lacked the cosmological sophistication to truly understand what was going on. Two: Alignment of astrological bodies might not be a very robust construct and has minimal influence on behavior. (We would tend to favor the second explanation.)

On a more serious note, we mentioned Freudian theory in Section 1.7 and the difficulty of **operationalizing** the concepts and **constructs** that make up that theory. Indeed, it might be easier to operationalize the cosmos than Freudian theory.

There are other constructs that are just not very robust, or so mediated by other factors that they are very difficult to research cleanly. For example, the notion of internal and external locus of control (Rotter, 1990) is not a very robust construct. It is mediated heavily by environmental conditions. An internally controlled person might be very externally controlled when in an unfamiliar environmental situation (say, an opera in New York City) and less so in a familiar setting (say, sidestepping cow pies at his farm in Kansas). Likewise, birth order effects (Schachter, 1963) are mediated by not just the order of births, but number of siblings, gender sequence in the sibling-line positions, distance in age between siblings, first-born/last-born status, recombined families, etc.

To illustrate further, with a less-complicated example, suppose we want to investigate and measure a construct like love. How do we define *love*? Certainly how we define it will determine, to some extent, how we measure it. We can be very pragmatic and define *love* as the number of years a couple has been married, but several of your authors have seen couples in clinical practice who have been married for a long time and are anything but in love with each other, or, if they are, it is a very peculiar form of love. We can take a more poetic stance and try to operationalize notions such as butterflies in the stomach, head spinning, or being swept off one's feet. That might be very hard to do.

How we define *love* makes a difference in how we measure love: married 26 years versus head-spinning rotations of 60 spins per minute. Can you begin to see the difficulty with some concepts and constructs?

Other constructs are easily defined because somebody else has done it for us. Depression can be defined by the Beck Depression Inventory. Gravitational force has been defined if we are dealing with rocket acceleration to infinity and beyond. Masculinity-femininity can be defined by Scale 5 of the Minnesota Multiphasic Personality Inventory (MMPI).

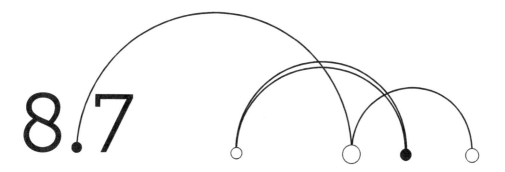

8.7

Measurement constraints

With any measurement, one must be aware of the influence of extraneous variables that could turn into confounding variables. We covered this in Section 4.7, so we will not belabor it here. Do realize, though, that there are statistical controls for extraneous variables, so extraneous variables will fit here and in Section 8.8.

With any quantitative measurement, the researcher must be aware of possible restriction of the range of the measure. An example will illustrate this. Each semester, students at one of the institutions your authors hail from complete a teaching survey on each instructor for each course. The teaching survey uses a five-point **Likert Scale**, ranging from strongly disagree to strongly agree. As a former department head, one of your authors noticed that most ratings given by students ranged from 4 to 5—very few 3s and no 1s or 2s. That means that a five-point scale essentially functioned as a three-point scale. It represents a restriction in the range of the scale that may reduce variance between the faculty, making them look more heterogeneous as teachers. A loss of variance can affect your data analysis and results.

Sometimes, when designing a study, researchers choose weak independent or dependent variables. If the **independent variable** is weak, it will be less effective in differentiating between groups. If the **dependent variable** is weak, it may not respond to manipulations in the independent variable. If both are weak, you may end up with nothing to show for your efforts. The best way to avoid this is to refer to your literature review to see what others have used successfully, and unsuccessfully, as independent and dependent variables.

One of your authors worked for a number of years as senior staff psychologist in a hospital for inpatient treatment of children and adolescents. We frequently administered the Children's Depression Inventory (CDI) to new admissions. One difficulty with the CDI is its **transparency**. It is a forced-choice format between two statements—choose the statement most like you now. One of the choices is pathological (for example, I feel like committing suicide) and the other is not (for example, I do not feel like committing suicide). Children and adolescents could easily manipulate the scale. Because of that, results were sometimes equivocal or did not match clinical interview data. The Minnesota Multiphasic Personality Inventory (MMPI) has fake-good and fake-bad scales to potentially identify individuals trying to manipulate response outcome.

There are also issues in marking and scoring. This is probably not a good example to mention, but many, many years ago, one of your authors decided to make grading a multiple-choice exam easy by making B the correct response to all questions. It seemed like a good idea at the time (this was before Scantron, so each test had to be scored by hand). Students taking the test felt for sure something was wrong because all answers could not be B, based on their past experience, so they started going back through, second-guessing themselves and changing answers. This was an error in instructor judgment—the exam scores were thrown out, and never has this instructor made all answers B again. He switched to C (just kidding).

Lastly, whatever instrument you are using could have an internal measurement error, meaning the instrument contains an error in its construction or calibration. One of your authors has a brother who is a physician. His brother often rails about failures to keep x-ray equipment properly adjusted. When a machine gets out of adjustment (and who would know just by looking?), patients may be exposed to more radiation than necessary.

The Stanford-Binet fourth-edition intelligence test is an excellent example of this. For a considerable time after the fourth edition was introduced, errata sheets were sent out correcting various errors in the tables, examples, scoring, etc. In addition, some of the directions for portions of the test were not very clear and psychologists were basically making their own interpretations of those directions, given the vacuum, and acting accordingly—and differently— based on their interpretations. This certainly presented some serious constraints on the IQ data generated by the test.

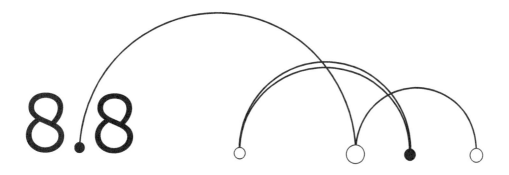

Error as error

All data theoretically have error. World time is kept accurate, for example, by an atomic clock. The atomic clock time is regulated by the emissions from a cesium atom and is accurate to 2 nanoseconds per day or 1 second every 1.4 million years. That it is off by 1 second every million or so years suggests that there is some degree of error even in the atomic clock (not that any of us need concern ourselves with that amount of error in our lifetimes). How much more error might there be when measuring more-amorphous constructs, such as intelligence; personality variables such as introversion or extroversion; learning curves for male and female pilots-in-training; or fuel efficiency of different airplane engines?

Even something as precise-appearing as measuring an individual's height is potentially fraught with error. Have 30 or so of your friends or classmates measure someone's height, record their results on paper without revealing them to others, and then look at the 30 answers. There is variation everywhere, and in that variation lurks the potential for error.

A few years ago, one of your authors took a civil engineering class as part of a personal program of neurological reinvigoration. This was in the days before computerized laser transits that automatically give distance and bearing, so we had to measure by hand, using steel tapes. The tapes had to be stretched under specified tension to minimize error caused by sag in the tape and, amazingly enough, the amount of expansion/contraction of the steel tape, given air temperature, had to be calculated mathematically and entered into the distance measured, again to compensate for error. As we said, there is error everywhere.

Error, obviously, could affect the veracity of any research results. In addition, there are so many sources of error. What we have discussed in the sections preceding this one all represent error in some form or another. When your author made all answers B, that was a measurement error, but also an error in judgment, so a "researcher" error as well. When our group on the WISC turned out to be significantly higher on grade point average, that was a random sampling error but also an error in the design of the study that did not account for that possibility. When a researcher does not follow the standardized directions for administering a measurement instrument, in essence creating his or her own idiosyncratic measurement instrument (as opposed to nomothetic measurements, which are norm-based comparisons), that researcher is creating measurement error, but that error is also researcher error because the researcher failed to follow appropriate protocols.

We can talk about error in a broader sense, by looking at **random error**, **systematic error**, and **errors of omission** or **errors of commission**. Let us start with random error.

In the example of the sixth-grade study of the WISC subtest, random assignment let us down. It did not distribute the quality of intelligence, as captured in academic ability through grades, across our three groups in an equitable manner. Our study received a severe blow. We were undone by random sampling error.

If you think about slot machines in Las Vegas, paying off would be considered a random error. The machines are designed to make money—the payoff is on an interval ratio schedule that approximates a random schedule, so putting coins in a slot machine is predicated on the hope that a random error—that is, a payoff—will occur.

We make systematic errors when we make a mistake on a repeated basis that influences the accuracy of the measurement we are making. Again, perhaps an illustration is needed.

In the civil engineering surveying course mentioned earlier, the textbook by Herubin (1983) stated, "Systematic errors are generally caused by imperfections in the manufacturing of equipment—not mistakes in the manufacturing process, but an inability to achieve absolute perfection" (p. 27).

The surveying equipment used back then required some degree of calibration or other adjustment each time it was used. For example, the transit plane had to be level. How much error might be introduced if the plane was not "perfectly" level due to an improper adjustment in the leveling screws? The transit also had to be oriented properly. What would happen if it were off by even a miniscule amount? Perhaps in surveying land, a few inches here or there may not make that much difference when looking at areas as large as an acre or more. What farmer might notice that his boundary is off by an inch? However, when looking at small tracts of land (for example, the deed to 1 square inch of Alaska that one of your authors received in his cereal box too many years ago to mention), an inch-worth of error could wipe out that holding.

We have already mentioned the atomic clock. The error associated with the cesium atom would be another example of systematic error.

In many fields, equipment and computers are often used to measure events or phenomena in one form or another. To the extent that calibrations are off or certain response sets have not been anticipated by the computer programmers, systematic error can be introduced into the data. The same possibility applies to computer-generated scoring systems for tests, such as IQ tests or achievement tests. If there is an error internal to the scoring program, then systematic error can be a result of each use of such scoring programs unless it is caught and corrected. Suppose one of the robot machines at some car assembly plant is off just a bit when installing axles. Oops! Not a car any of us would be interested in buying! Until caught, this error would perpetuate with every car coming down the assembly line.

Errors of omission, meaning something has been left out, *might* also be termed "errors of ignorance" in that they are generally created when an individual collecting data does something wrong but is not aware of doing so. (Errors of ignorance do not imply that the individual is ignorant; rather, just unaware or improperly trained.)

For example, one of our students a number of years ago worked at a psychiatric clinic as a psychometrician, primarily giving IQ tests to adults. The student claimed that he had been thoroughly trained on such tests and asked to be given credit for the IQ testing course based on his training and experience. The student was asked to administer an IQ test while being observed by the IQ test course instructor. After the observed administration, the student was asked to critique his administration and responded that everything had gone well and the results of the test could be considered valid.

When the instructor reviewed and commented on the administration, pointing out numerous mistakes, it became apparent that the student was making those mistakes out of ignorance and lack of familiarity with test administration instructions. The student's administration was a collection of bad habits that had never been corrected.

Most of the tests that student administered at the clinic were being used in a large database for a study. Imagine how chagrined the researchers would have been to know that so many mistakes had crept into their database. Once the student knew better (after taking the course—no exemption was granted), his test administrations improved considerably. His errors were errors of omission—in this case, his "ignorance" of proper administration procedures.

Particularly in applied fields, it is easy for errors associated with habit, forgetfulness, or insufficient training to sneak into results from such fields. This is one of the reasons why it is good to periodically refresh your knowledge of familiar techniques through a training workshop, observation by a trusted colleague, or other means of review.

Errors of commission occur when the individual knows better regarding some procedure but deliberately chooses to deviate from accepted practices. An example might clarify this. An acquaintance of one of your authors from a number of years ago did contract work with the Bureau of Disabilities of a local municipality. This contract work required administration of a test battery that had to include an accepted IQ measure, as well as the Rorschach Inkblot Test. This individual administered short forms of both tests. That is, he dropped one subtest from the Verbal and what was then the Performance section of the Wechsler Adult Intelligence Scale, and gave only the odd-numbered Rorschach cards.

While much research has been conducted on short forms of the Wechsler scales, they are not used now due to questionable psychometric properties, and, to the best of your authors' knowledge, nobody has advocated for a short form of the Rorschach.

This particular individual knew better. He simply chose to operate in this way because it allowed him to conduct more assessments than otherwise. Back in those days, several of us frequently collected data from colleagues for various research studies, often dealing with assessment results. We steadfastly avoided ever using any data from this particular individual.

The wife of one of your authors, when a teenager, had a car and knew she was supposed to check the oil and change it regularly. She simply did not do it—too much else going on. When her engine seized from lack of oil, she realized the error of not doing what you know needs to be done, particularly after her father refused to fix the engine or buy her another car.

We hope you can see that error is a multifaceted consideration in research. Error, as we have defined it here, is certainly not limited to psychology or psychological research. It applies to all fields that conduct research, perhaps even more to those fields that rely heavily on instrumentation of a sensitive nature that must be calibrated or operated properly.

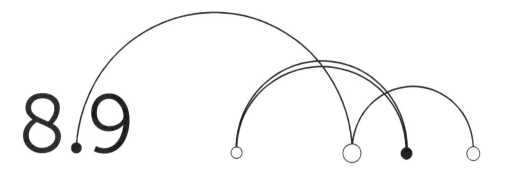

Statistical error

A number of years ago, one of your authors returned to his doctoral institution and looked up his dissertation chair just to say hello and thank him again for his patience those many years ago. His dissertation chair was excited about the findings of a doctoral student he was working with at the time. In what he thought was a groundbreaking study, they had found nearly perfect group discrimination using discriminant function analysis. When he was asked about the study, it quickly became apparent that the number of variables under examination exceeded the number of subjects. In a discriminant function, this almost ensures very high group discrimination, a function of the statistical technique rather than the data (Tabachnick & Fidell, 1989). What a bummer!

This illustrates a point, however. We trust to those who write analysis programs that what is going on internally to the program is correct and, therefore, our data are trustworthy. Many—nay, most—of us do not have the expertise to delve into the internal commands of SPSS, SAS, or other analysis programs to check the internal efficacy of what has been written. We are at the mercy of those who write the programs.

This applies to just about any field. Architects use software to lay out designs, foundations, homes, etc. Engineers use transits, now with built-in computers, to get distances, angles, and orientation. Think of the amount of computerization that had to go on to get your last MRI. We certainly trust that all of this has been programmed correctly and will not give us false results.

There are other statistical constraints. Many deal with the expertise of the researcher to design, collect, and analyze data by choosing the correct analysis procedures. Failure to choose correctly can lead to statistical errors secondary to researcher constraints.

There is, of course, Type I and Type II error, which you probably covered in your statistics course. Sampling error can fit under statistical error.

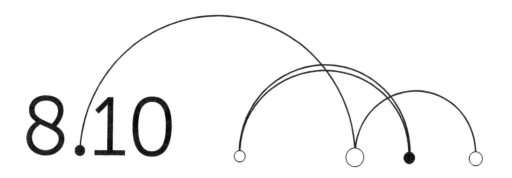

8.10

Guarding your research against threats

For research to be good, apart from issues of relevance, etc., you must guard against these constraints and potential errors. One way to do that is through careful review of the literature to see where others may have run into problems, and to be sure that you are on a solid track with your own study. Another way to do this is with thoughtful and careful planning as you develop your research.

Ask yourself relevant questions such as: Is this research question the next step in the perpetuation of this field of study? Can I operationalize my variables? Is there a relevant source of data for this study? Do I understand the analysis methods I have selected? Can I interpret the data I generate? These basic questions must be asked. A research study should be reviewed at every step in the development of and progression to data collection and analysis.

Careful design of your study is important. Keeping extraneous variables and other threats at bay will be enhanced if your research is taking place under laboratory conditions. In the lab, you can set schedules, control environmental factors (temperature, light, sound), use precision measurement devices, ensure that each participant receives the same directions, etc. Labs simply provide an environment over which you, as the researcher, have a great deal of control. It aids uniformity of procedure at each step along the way to a finished study.

Sometimes research does not take place in a lab, though. In the real world, many of those aspects of lab conditions that you would otherwise have control over go out the window. That is not to say that research in the real world cannot be conducted. It does mean that you have to be that much more attentive to potential threats to the rigor and accuracy of your methods.

There are some things you can do in any situation, but particularly in the field, that can help. Foremost is random selection and/or random assignment of participants. This eliminates a plethora of possible threats and extraneous variables. Second, you can use assessment procedures that are reliable and valid, based on the properties of the procedure or past research using the procedure. You can also make sure that the independent variable is robust and that the dependent variable is responsive to manipulation of the independent variable.

There are also statistical means of controlling extraneous variables and threats apart from those already mentioned. You can use **counterbalancing** to offset test-retest effects. As an example, you might administer test A to half the subjects first, followed by test B. The other half of the subjects would get test B first, followed by test A. The counterbalance negates any test-retest effect.

You can also conduct **single-blind** or **double-blind studies** to reduce any biases or predispositions of subjects or data collectors.

In a single-blind study, the subjects would not know the intent of the study, so the subjects are "blind" to the intent of or reason for the study. Therefore, their behavior should not change based on knowledge about the study. Likewise, in some causes, you may not want the raters or note-takers to know the intent of the study, in case they become predisposed to rate in line with researcher expectations (i.e., the raters are blind to the intent of the study).

When both participants and raters are blind, it is called a *double-blind study*. You can also use unobtrusive measures to avoid changing participant behavior based on participants perceiving they are being observed.

Another approach would be to use a pilot study. A pilot study is a mini-version of the real thing. The researcher collects enough data to make sure everything is working properly and being measured accurately, and that there are no undue influences on the study that might call the results into question. Pilot studies are most relevant for longitudinal research and expensive research. You do not want to invest 10 or 15 years, or thousands or millions of dollars, in a study only to realize at the end that the data were fatally flawed in one or more ways that could have been corrected easily if only the researcher had known.

Key Terms

- Concepts
- Confounding variable
- Construct constraints
- Constructs
- Counterbalancing
- Demand characteristics
- Dependent variable
- Double-blind study
- Environmental constraints
- Error constraints
- Error of commission
- Error of omission

- Extraneous variables
- Impression management
- Independent variable
- Likert Scale
- Measurement constraints
- Operationalization
- Participant constraints
- Pilot study
- Random assignment
- Random error
- Reactivity
- Recording

- Reliability
- Researcher constraints
- Single-blind study
- Socially appropriate direction
- Statistical constraints
- Systematic error
- Transcribing
- Transparency
- Validity

Questions

1. What differentiates a single-blind study from a double-blind study?

2. You are studying error rates by gender for workers on an aircraft assembly line. Identify three extraneous variables that will have to be controlled.

3. A researcher is running some multivariate analyses on data. Not being trained in multivariate techniques, he has developed these options: Consult with a multivariate expert, take a course, or read up on the subject. Which option is best and why?

4. A researcher hires an undergraduate student to code some data for him. The undergraduate is not familiar with the study, data, or process of coding. What should the researcher do?

5. Give an example of reactivity from your own experiences.

6. Define and operationalize honesty.

7. Create three questions measured with a Likert Scale to obtain data on employee workplace satisfaction.

8. Give an example of either random error, systematic error, or errors of omission or commission from your own experience. Explain.

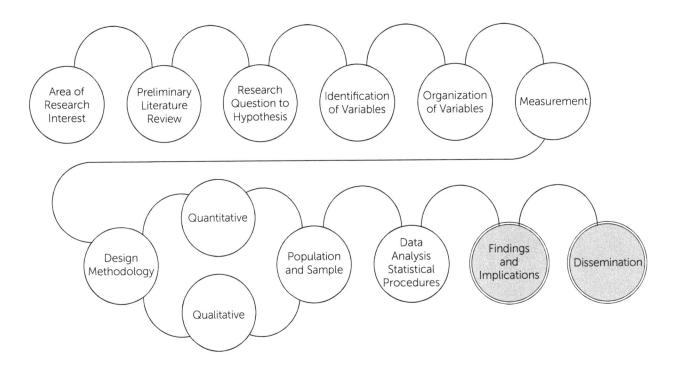

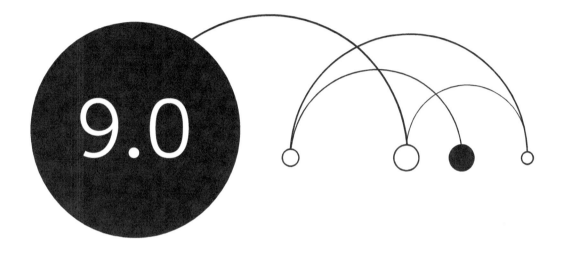

Ethics and Cultural Issues

The professor of research always tells his students, "Remember, *data* is always plural."

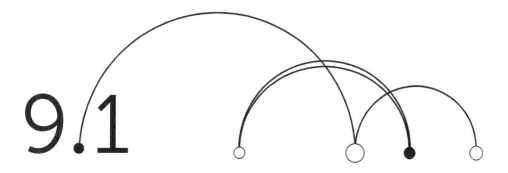

9.1

Ethical limitations

While the conduct of research today is subject to stringent controls and safeguards, especially with regard to the use of human subjects, this hasn't always been the case. In the past (and sometimes not-too-distant past), research, particularly that of a medical nature, took more of an "ends justify the means" perspective over strict adherence to an ethical code or a concern for the well-being of human subjects.

Throughout the 1940s, 1950s, 1960s, and 1970s, various agencies—both public and private—carried out potentially dangerous experiments, intentionally exposing vulnerable and/or unknowing human subjects to risky conditions, including viruses and bacteria, live cancer cells, biological and chemical warfare agents, and radiation. Included in this infamous group were, among others:

- The Stateville Penitentiary Malaria Study
- Operation Sea-Spray (1950)
- Willowbrook State School "vaccinations" (1950s –1972)
- Jewish Chronic Disease Hospital, Brooklyn, New York (1963)
- Project Shipboard Hazard and Defense (SHAD) (1963–1969)
- A Study of the Vulnerability of Subway Passengers in New York City to Covert Attack with Biological Agents (1966)
- Tuskegee Syphilis Experiment (U.S. Public Health Service, 1932–1972)

Many of these experiments were performed on children, the sick or infirm, the mentally disabled, or prisoners, who could not decline participation themselves and who did not have advocates protecting their best interests. In addition, various aspects of the experiments were either hidden or misrepresented to both the subjects and the public.

Beyond the examples in the medical field, several iconic studies in social psychology represent examples of questionable ethical practices with respect to the psychological well being of participants. These include the Robber's Cave Study (1954), Milgram's Obedience Experiment (1963), the Tearoom Trade Study (1970), the Blue Eyes/Brown Eyes Exercise (1970), and the Stanford Prison Experiment (1971). While each experiment yielded interesting insights into human behavior and interaction, they also generated controversy about methodology and manipulation, and most would not be allowed to proceed (at least in the same form) in today's more tightly controlled research environment.

In 1972, a leak to the American press shut down the decades-long Tuskegee Experiment and led to enactment of the National Research Act (1974), which established the National Commission for the Protection of Human Subjects of Biomedical and Behavioral Research. The commission was charged with identifying "the basic ethical principles that should underlie the conduct of biomedical and behavioral research involving human subjects and to develop guidelines which should be followed to assure that such research is conducted in accordance with those principles." Specifically, in this context, the commission investigated (1) the boundaries between biomedical and behavioral research and the accepted and routine practice of medicine; (2) the role of assessment of risk-benefit criteria in the determination of the appropriateness of research involving human subjects; (3) appropriate guidelines for the selection of human subjects for participation in such research; and (4) the nature and definition of informed consent in various research settings (National Commission for the Protection of Human Subjects of Biomedical and Behavioral Research, 1979, p. 1).

In 1979, the commission issued The Belmont Report: Ethical Principles and Guidelines for the Protection of Human Subjects of Research, which detailed the foundational principles of respect for persons, beneficence, and justice that are necessary for use in research with human subjects. These principles form the basis of ethical codes in practice today and the criteria upon which Institutional Review Boards (IRBs; see next section) base their decisions and oversee the conduct of research.

Respect for persons refers to the fundamental right of individuals to act freely and autonomously. This implies that subjects, or potential subjects, in an experiment should be provided the opportunity to act in their own best interests, particularly with respect to decisions related to participation. It also highlights the need to protect special populations (such as children, the mentally disabled, and prisoners, as noted above) who may have "diminished decision-making capacity" and limited autonomy. In short, individuals should not be subjected to coercion or undue influence to become subjects in a research study.

Beneficence follows the edict of "do no harm." It requires the researcher to maximize potential benefits while minimizing potential risks. This principle underlies every part of the research process, from choosing an appropriate, and valuable, research question through the selection of subjects, measures, and methods, to the proper handling of data and dissemination of results. At each juncture and with each decision, researchers must consider the possible consequences, good and bad, of their research, and the effects on both individuals and society at large.

Justice, in this context, refers to a fair and equitable distribution of the burdens and benefits of research. It relates primarily to the selection of subjects so the poor, disadvantaged, or underprivileged are not exploited or marginalized. Many studies of the past used convenience samples of captive populations without ensuring their fair treatment (e.g., the Tuskegee Syphilis Experiment).

Researchers are typically enthusiastic—and sometimes singularly focused—when it comes to pursuing their lines of inquiry. For that reason, despite their good intentions, they may lose perspective and fail to anticipate the consequences and risks of their activities. Controls such as peer review and IRBs function to ensure a safe and just research environment for all.

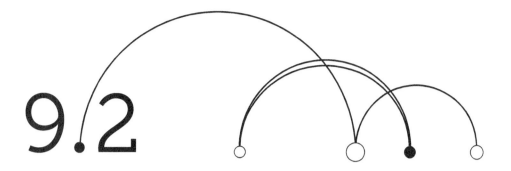

9.2

Research oversight: The IRB

Institutional Research Review

The U.S. federal government has established guidelines for using humans in research. These guidelines may be found in the *Code of Federal Regulations:* Protection of Human Subjects, 45 *CFR* 46. Most institutions that conduct research have boards that review such research for compliance. The Institutional Review Board (IRB) scrutinizes all research proposals that deal with research involving human subjects to review ethical circumstances. A researcher (typically including student projects) who intends to use human subjects must submit an application to the IRB for review before conducting any research. Failure to comply could cost an institution or researcher any federally supported research grants or federal support of other kinds.

The essence of the guidelines is protecting subjects from harm that might be caused by deception or physical intrusion. These guidelines would most certainly have prevented J. B. Watson's study with Little Albert, had they been in effect at that time, and perhaps seriously hampered Stanley Milgram's studies at Yale. (If these two studies are unfamiliar to you, you might want to look them up. Both are interesting reading.)

At the heart of protection from harm is the notion of informed consent. Simply stated, this means that participants need to know what is going on and what will happen to them as a result of the study unless such disclosure will potentially distort responses. For example, if you were to conduct a study looking at the extent to which people can be considerate under specified conditions and you told participants the purpose of the study, then you just primed them to act considerately rather than display normal reactions under the specified conditions.

Informed consent, at a minimum, involves the following:

- An indication that the study involves research and the amount of participant's time that will be needed to participate;
- A statement that participation is voluntary and the participant can refuse to participate or stop participating at any time without bias or penalty;
- A description of the procedures to be used, including disclosure of any risks to the participant;
- A description of the confidentiality of any records or record storage, and whether analyses will be individual or aggregate;
- A statement of any compensation for participation.

Additional guidelines become activated as the potential risk or intrusiveness of the research increases. Any modifications to the guidelines must be approved by the IRB and must be justified based on the study requirements. For example, any deception of participants must be fully explained in the IRB submission to gain approval. Table 10.1 provides a brief list of examples of the type(s) of risk that participants may encounter and what type(s) of special action may be needed.

TABLE 9.1 Assessment of Risk

Risk Assessment	Examples	Special Actions
Exempt research: no risk	Anonymous surveys, observations of nonsensitive public behaviors where participants cannot be identified	No informed consent needed, but must be reviewed by IRB
Minimal research risk	Standard psychological measures, voice recording not involving danger to participants	Informed consent may be needed
Greater than minimal risk	Research involving physical or psychological stress, or invasion of privacy	Full IRB review required, special ethical procedures may be needed

Source: *Methods in Behavioral Research*, P. C. Cozby (2009).

This has been a very brief overview of the IRB process and requirements. If you have any questions about what is required, make sure to check with your institution's review board.

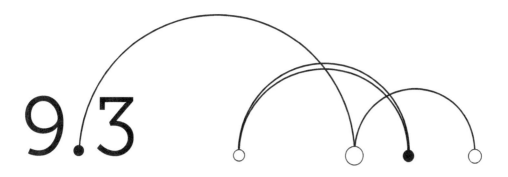

Cultural limitations

In 1884, an article appeared in the *New York Times* suggesting that **slumming**, already seen in England, would become the rage in New York and other large American cities (Heap, 2010). What is slumming? It involved upper-class whites touring through urban ghettos to observe and marvel at the lives and behaviors of immigrants. The most-popular locations for slumming were Irish, Chinese, and Italian neighborhoods.

Slumming is, at heart, an example of **ethnocentrism**. Ethnocentrism is the tendency for all of us to look at other cultural, ethnic, and racial groups through the lenses of our own cultures. Such ethnocentric myopia can lead to misunderstandings, inadvertent insults, tensions, conflicts, and mistrust. We see it currently each day in the evening news. Stepping away from that myopia usually takes major and concentrated effort. Even with effort, one could raise the question of whether our cultural lenses will ever be neutral toward other cultures, no matter how hard we try. The first step, however, is awareness.

As a budding researcher, perhaps one involved in research across different cultures (**cross-cultural research**), you do not want to be myopic. You want to see clearly the behaviors you are examining across those cultures so, ideally, you can determine which are universal (often thought to be due to nature, or genetics) and which are limited and culturally specific (often thought to be due to nurture, or environment). Indeed, identifying the universal from the not universal is at the heart of cross-cultural research (Matsumoto, 1994).

Why is this research important? For the very reason that the **nature-nurture** debate has been going on for centuries: so we can have an understanding of those universal behaviors, assumed to have a biogenetic base, and behaviors that are learned from our environment. The learned behaviors, supposedly, can be modified. Contrary to earlier beliefs, even the behaviors that are biogenetically based can be modified. We are not slaves to our genes psychologically!

Here is an example: At one time, Down syndrome children were almost automatically given up at birth and housed in state mental health facilities. The triplicate chromosome resulted in mental retardation and that genetically based condition was assumed to be immutable. Indeed, the degree of retardation evidenced after years in a sterile institutional environment simply confirmed the initial decision to institutionalize. Now, such children are raised at home with support services and enriched environments, and the degree of retardation can be modified by such an environment, up to a limit.

If interested, the reader is directed to the studies of Ekman (1999) as an example of universal versus specific behaviors, particularly with the Fore Tribe of New Guinea. Ekman identified specific facial expressions that he initially determined were universal in nature—that is, every culture recognized these facial expressions. This universality would suggest a common cross-cultural ability ingrained in all humans. It is an interesting area of research to pursue just for one's own edification.

We will approach cultural influences on research by looking at several specific cultural differences that have been highlighted by research as examples of how those differences might influence research outcomes. Before we do this, though, we have to accept that when we talk about culture, we are including the social, economic, religious, and psychological structures that are embedded in cultures. We are also accepting that geography plays a part in cultural differences. These are givens. Now, on to specifics!

Response Sets

A response set is a tendency to answer in a certain direction. For example, in the many studies that have been conducted on sexual behaviors among teenagers and adults, there is the assumption that people tend to answer in socially appropriate directions. This is a response set.

What kind of response sets do we see culturally? Research has suggested differences among cultures in the tendency to disclose personal information, individualism versus interdependency, process of making causal attributions (attribution theory), tolerance of dissonance, figure-ground discrimination, and subjective sense of well-being, just to name a few (Chiu, 1972; Diener, Oishi, & Lucas, 2003; Ji, Nisbett, & Yonjie, 2001; Mesquita & Frijda, 1992).

An investigation of self-esteem cross-culturally between the East and West might well pit Western individualism against Eastern interpersonalism. An investigation of problem-solving approaches and abilities might pit Western analytical thought against Eastern holistic thought. Given these opposing response sets, how do we separate what is cultural from what might be behavioral or neurological? And how does one make sense of findings? It becomes very difficult to do.

For example, let us propose this research question: Are mathematical abilities greater among individuals of Asian descent than among individuals of Caucasian descent? To test this, we randomly select a group of exchange students from China who are studying in the U.S. and another group of native-born Caucasians. We administer some math test that has proven to have reliability and validity in past studies and has proven to have discriminatory capacity in native math skills.

Let us now suppose we find a difference, with our Chinese exchange students scoring significantly higher than the Caucasians. We now write a conclusion as follows: "These results suggest that Caucasian children need to be taught math skills starting at an earlier age, with more-intensive instruction as they age." That seems a reasonable conclusion, wouldn't you say? In response, we start a program at an early age, increase intensity, follow students longitudinally from this program, and, just before we retire, do another comparison with Chinese exchange students. We find the same significant difference in favor of the exchange students is still there. Where did we go wrong?

It could be that the difference is not in the basic skills, but rather in the language (analytic versus holistic) that allows a different linguistic conceptualization of quantities for the exchange students that is not possible given the language structure of English.

This is a made-up example. The point is that differences might not be overcome in some instances by training since they are inherent in the cultural structure the individual hails from.

Research Constructs

What we are investigating—our research area of interest and subsequent research questions—may well be influenced by cultural differences. In 1947, Alfred Kinsey began investigating sexual behavior of males and females at the Indiana University (Bloomington) Institute for Sex Research (readers are directed to the 2004 film *Kinsey*, starring Liam Neeson). Even then, the research generated considerable controversy and opposition. Imagine carrying out that same research in areas of the world where sexual behaviors are governed by cultural and/or religious constraints.

One of the reasons why sex figured so pervasively in Freud's theory was the sexual repression of Vienna during the late 1800s and early 1900s (Gay, 1988). Would it make sense for you to design a study to look at computer literacy comparing American youth with age-appropriate counterparts belonging to the !Kung tribe of the Kalahari Desert? That study would never get off the ground. How about a study looking at the extent of self-actualization based on Maslow's hierarchy of needs between Americans and Somalians? That is another nonstarter!

Not only are some of these nonstarters just because of the topic area, but one also has to determine the equivalence of terms between cultures. For example, does self-esteem mean the same thing to Americans as to a seemingly similar group, such as the Irish, or a more dissimilar group, such as the Japanese? Do Americans define achievement in terms of our individualism while another culture defines achievement in group or tribal terms? If there are cultural differences in definitions, then—of course—there should be group differences in the construct.

In thinking about construct equivalence, we can bring quantitative and qualitative techniques into play. Quantitatively, you might use an existing instrument to assess self-esteem, such as the Coppersmith. Would this instrument be appropriate for another culture?

Perhaps the question of appropriateness should be answered first. This might involve a qualitative exploration where the aim is to determine difference in the equivalence of the term across the cultures being investigated. Then, with a clearer notion of the cultural influence, more quantitative approaches might be brought to bear. This could be an interesting mixed-methods study (see Figure 1.6 from Chapter 1).

Language Influences

You have a really good idea for a study using a questionnaire. You have addressed issues of construct equivalence, etc., and have written your questionnaire. Now you must get it translated into Luxembourgish. Can all the terms in your questionnaire translate directly? Are their obvious equivalent words in the other language? Will you lose meaning in the translation or end up with misperceptions about certain questions?

This is not a trivial matter. Differences in cultural response based on the inadequacy of a translation, rather than true cultural differences, are basically garbage data. A technique called back translation is often used: The instrument is first translated into the other language, then translated back into the original language to see if it comes out close to the original instrument, usually with the second translation being made by another expert in both languages.

Even with the possibility of accurate translation, some languages influence cognitive and conceptual process simply because of the way they influence and organize thinking, as alluded to above in our math differences study. For example, Franz Boas (1922) found that the Inuit had 50 different words for "snow." Let us suppose you are standing on the corner waiting for a bus, hoping that it will indeed come by in the terrible New England snowstorm you are having. An Inuit comes up to the bus stop as well, and the two of you begin to discuss the weather. Will you ever be at a disadvantage!

For an excellent exploration of cultural differences, particularly as they relate to language, we recommend Barry Lopez's (1986) telling account of his life among the Inuit. This is an excellent read and explores issues such as prosocial behaviors, environmental awareness, etc., that are a function of the history of a culture and the geographic region where the culture developed and exists.

Languages are complex and involve orientation frames that differ across cultures, e.g., analytical versus holistic, concrete versus abstract, functional versus symbolic, etc. The language you use is an integral part of the cultural lens you apply to other cultures. It influences your frame of reference, your cognitive processes, your expression of your humanity, your conception of your relationships to others, your expression of your psychological wellbeing, etc. Language is a pervasive and often subtle part of that cultural lens we all look through.

Geographical and Environmental Influences on Culture

Your first reaction might be that we have to be kidding—how can geography influence culture? Let us explain. A number of years ago, two of your authors piled into an old Volvo 240 and spent three weeks touring Europe. For one author, who lived in Europe, this was familiar territory. For the other, who was there for the first time, it was all new. One of the enduring memories of that trip was the depth of the history and length of people's timeframes, particularly compared to the United States, then only a little over 200 years old.

To put this in blunt perspective, the Citadel was established in 1842. Two years ago, two of your authors presented research at a conference held at the University of Groningen, established in 1614! That history stretching so far back influences time-binding, and time-binding is a cultural trait.

Another example: Segall, Campbell, and Herskovits (1966) examined cultural differences in the Muller-Lyer (look it up) illusion. Would you believe they found differences between Plains Indians and urban dwellers? The Southwest of this country is composed primarily of horizontal vistas (plateaus, horizons, etc.) and urban areas primarily of vertical vistas (skyscrapers, towers, light poles, etc.). Where you grow up has something to do with perceptual processing, and perceptual processing is integrated into cultural perspectives.

Geography has biological influences as well. Around the equator, people tend to be thin and linear—built to dissipate heat. The Inuit are round and short—built to conserve heat. That biological and environmental difference translates into cultural differences in clothing styles and daily patterns, e.g., late dinners in hot climates when the air has cooled, early dinners in northern climates when the sun has dipped and the air has lost any heat it may have had.

Methodological Influences

Some quantitative research techniques make use of random samples in an effort to find group similarities or differences. Random sampling, we hope, distributes extraneous variables evenly and normally across two random samples drawn from the same population. There is, however, the assumption of some degree of homogeneity in the population, and thus, the samples. Would this be true with culture? If you randomly draw a sample of U.S. citizens from California and another random sample of U.S. citizens from New York, will they be similar across all variables? Some, probably yes—like age, gender, socio-economic status (SES)—those more cut-and-dried interval/ratio variables or biologically determined nominal variables.

But how about culture? We do not have an answer to that question, but it is certainly something to contemplate. The assumption tends to be that people in a definable biome are alike in culture. We are not at all sure that is a valid assumption. If one begins to do cross-cultural research, the issue becomes even more problematic. If we compare American to Japanese, can we assume that all the people we sample in Japan are Japanese or that all the Americans meet our perhaps-stereotypic notion of what Americans are like? Can we make assumptions of cultural homogeneity? Can we make that assumption anywhere, given the mobility of the world population today?

When we make cross-cultural comparisons of behaviors, cognitions, etc., can we filter out that which is cultural? Again, we have no answer. Moreover, if we have no answers to these methodological questions, what does that imply about interpretation of results? We would suggest it implies a good deal of thinking and conservativism in extrapolation from a sample to a population, or from one sample to another.

As an additional point; if we are looking at two different cultures for a difference, say on self-esteem, do we expect to find a difference? The obvious answer would be yes—they are two different cultures! Perhaps the more important research question might be not whether they are different, but *why* they are different.

Enter qualitative methods again. Cross-cultural research seems to lend itself very well to mixed-methods approaches as a way to try to get around some of the instrument, language, timeframe, and geographical/environmental issues already mentioned.

Final Thoughts

Culture is essential to human life. It grounds us. It contributes to our sense of belonging. It helps us define who we are and where we fit into this puzzle called life. It is neither abandoned nor changed easily at the individual level. The task is to see around the corners of one's own culture—not necessarily change one's culture—to other possibilities that are just as necessary and valid for others as your own culture is necessary and valid for you.

Where to Next?

You are basically at the end! In Chapter 10, we will provide a rehash of the research methods and processes covered in this book. This next chapter will be quick and easy, we promise.

Key Terms

- Beneficence
- Culture
- Ethics
- Informed consent
- Institutional Review Board (IRB)
- Justice
- Respect for persons
- Response sets

Questions

1. What ethical questions were raised by early medical research? Be specific and cite a particular study as an example.

2. What ethical questions were raised by early social psychology research? Be specific and cite a particular study as an example.

3. What was the significance of the Belmont Report?

4. What is meant by the terms respect for persons, beneficence, and justice, with respect to research ethics?

5. What is the purpose of an Institutional Review Board?

6. Why is informed consent important in research?

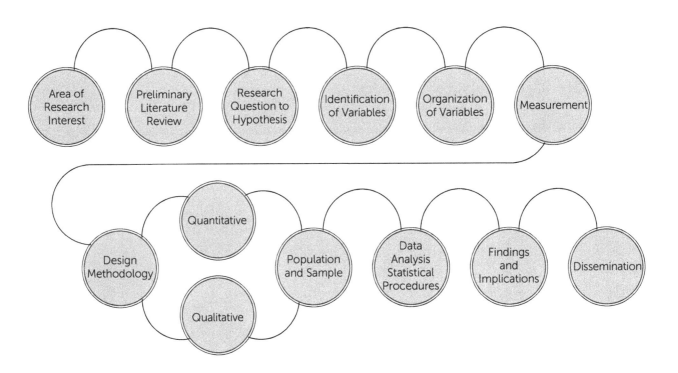

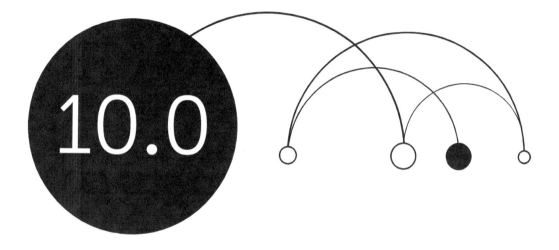

A Rehash

An absent-minded research professor is caught speeding while trying to figure out a problem that has been bothering him for months. "Let me see your driver's license," says the cop. "Okay," says the absent-minded researcher. "It has my picture on it, right?" "That's right," says the cop. The absent-minded researcher finds a small mirror, looks at it, sees his face, and hands the mirror to the cop. The cop looks at it and says, "I'm going to let you go. I didn't know you were a cop."

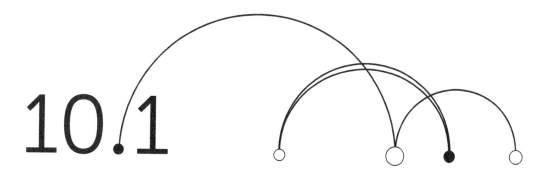

10.1

Why do we need a rehash/recap?

When your authors were doctoral students, they had to take written and oral exams—writtens first, then orals, so you could redeem yourself if you messed up on any portion of the writtens. At Indiana University, in psychology, you wrote blind for three days, meaning you did not know what questions you might get. All you knew was that you would write in the basic areas of learning, cognition, development, social, abnormal, statistics and research methodology, personality, and experimental-physiological.

Many students took the four to five months before the writtens to form study groups that met daily, often in the deep recesses of the library, for hour upon hour of review. In a group, each person would take one area, review that in-depth with notes for the others, and at some point "teach" his or her area to the others. At the time we started preparation, the general question was, "Why take writtens? We passed the courses!"

By the end, the answer was apparent: We had integrated across content areas that went well beyond classroom work. There were no longer separate areas of study, but rather overlapping and complimentary areas, each informing the other in a breadth and depth that extended well beyond what we learned in the individual courses. We realized the real purpose of the writtens was not to punish students, but to provide an opportunity for integration of the broad body of material we had been exposed to for the last five to seven years.

There is the reason why a rehash/recap is important: to solidify what you know and to integrate your knowledge into a broader frame than just this one course. One of the things we hope to do here is to show how, in effect, your life has been a research project in the making and parallels what we have covered in this book.

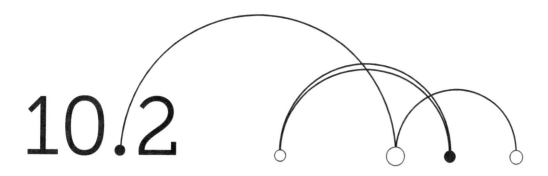

10.2

Takeaway points from the study of research methods

The first takeaway point is that research is a process. It is a process of expanding the boundaries of knowledge. It is a systematic process of building on what has gone before. It is a constant process of asking questions, seeking answers, and, based on those answers, being able to see questions that were not visible before because the knowledge you needed so you could see those questions was limited (this is the process of equilibration; see Piaget).

Each time you learn something new, the foundation is laid for you to see new vistas thanks to the new insights. We suggest it was hallowed ground and, really, no matter how pedestrian you might think your research study is, it is still hallowed ground. If you read through what was just written with a perspective not limited to research methods, you will note that we have also just described the process of life, and life is always hallowed!

Like research, life is a process—one of acquiring new knowledge, building on the old, forever seeing something new not glimpsed before, with the new driving you on to seek further, look harder, accomplish more, think deeper, strive, build, ask questions, etc. Did you take this course because you are looking ahead—have a goal in mind—something you are striving for? Most of what you do today has been built on the past, just like research builds on past research.

Think of your current relationships, singular or plural. Are they more sophisticated than when you were 4, 10, or 15? Did you learn from the process of negative results what modifications to make such that future results would be more significant? We think you have actually been doing research all of your life, testing one hypothesis after another, discarding the negative results, pursuing the ones that produced positive results. This course, we might suggest, and what you have learned is a recap of your life and how you have progressed from where you were to where you are now. As research goes, so goes life!

We also suggested that the driving force behind research is an area of interest (we hope) that leads to a research question. Why are you in this course? Are you perhaps pursuing an area of interest? If so, then, again, life mimics research—or would that be the other way around?

When you decided to pursue this area of interest, it was just a general thought: "I think I would like aeronautical engineering." That is a good thought, but it will not get you anywhere very fast. It has to be operationalized. You need to come to grips with the how and what of getting there. To do so, you start looking at college programs, those programs are composed of courses, and the courses are building blocks to the end goal of becoming an aeronautical engineer. Operationalized, your area of interest now is to enroll at XYZ University and take the undergraduate program in engineering. That would be a good operationalization of "I think I would like to be an aeronautical engineer."

While your general statement of interest is the broad driving force, the operationalized statement outlines the steps to get there and provides the necessary data to evaluate progress in the form of number of courses completed, grade point average, etc.—all good ratio variables that can be used to track progress.

There are key points along the way in research that determine how well your study will turn out. Good decisions lead to good outcomes; less-thoughtful decisions may lead to less-positive outcomes. (We would bet you have already seen the parallel to life in what we just said.)

One of those decision areas we have already mentioned is your area of interest. We have also mentioned the all-important research question.

A number of years ago, one of your authors was watching a TV interview program. One individual being interviewed was a sales representative for a life insurance company based in Georgia. The interviewer asked him if he had taken business or something similar as an undergraduate to prepare for life as a salesperson. "Oh, no," he said, "I was a premed undergraduate." The interviewer asked, "Why are you doing this instead of attending medical school?" "Because," he replied, "I got married, we had a child, and so I took this job." I have always wondered if he had really pursued his area of interest—if at some point (mid-life crisis?), he would regret the decision to persevere despite the wife and child. Was his hypothesis about his life a good hypothesis?

Another key point is the literature review. You are probably thinking this will be hard to stretch into our analogy we have been building—a literature review as a function of life? No problem!

Literature is your guide to evaluating the efficacy of your study: Has it already been done? Are there tweaks that have to be made for it to be more current and potentially important? You can do replication studies and such studies are all right. They solidify the existing knowledge base, even if—perhaps—they do not move it much beyond current boundaries. However, you do not want to replicate, exactly, your parents or siblings. Replication studies with new tweaks that address what has not been addressed before are even better.

We can see the benefit of these types of studies in the relationship of tobacco to cancer. Such an overwhelming mass of findings has been compiled that we now have warnings on tobacco products, age restrictions on sales, and constant advertisements depicting the potential outcome of continued tobacco use.

Another area that illustrates these types of studies is media violence and the relationship to decreases in pro-social behaviors. Again, the evidence is rather overwhelming. The continued depiction of violence in movies, television, and video games probably reflects the cultural transportability of violence and the massive amounts of money made from sales of media containing violence. For additional information about this conundrum of the continued production of violence, look to the work of George Gerbner at the Annenberg School of Communications of the University of Pennsylvania. These quasi-replication studies could be reflective of your life. We are sure you get the connection.

Studies that are not replications and really push against the boundary of knowledge are even better. It is your literature review that will give you the necessary information to decide where your study fits or what you need to do to your research question to push it harder up against that boundary. Such studies represent a greater break from the past—a bold move forward.

As we write those words, the movie *Breaking Away* (very enjoyable) comes to mind. You see an individual making a complete break from the past in an effort to carve a new destiny for himself—another life analogy connected to the types of studies we have covered in this book.

Students always seem to groan when asked to do a literature review, yet you have been doing them all of your life. Ever read a newspaper? Read a book? Listened to a teacher? Particularly listened to your parents and, most importantly, to their admonitions? There is your literature review of life! We imagine your parents were trying to keep you from replicating something: your past behavior, their lives, etc. They wanted you to read new material; learn new information; find that breakaway that fits their dreams for your life, if not your own dreams. Your reading through the years, along with conversations with teachers and significant others (including parents), has formed a general base of knowledge that has guided your life, whether you can pinpoint a particular life decision to a particular passage in Hemmingway or not. It forms the basis for an evaluation of the viability of your research question coming out of your area of interest.

The next step in research is to identify data collection techniques and sources. This is aided greatly by your literature review. It would indeed be near impossible without that literature review. What data from which sources have you collected to evaluate your current area of interest and resulting research question around which you have built your current life? Might we suggest income as an important ratio variable or anticipated income? Grade point averages? The significance level of relationships based on a Likert Scale? Satisfaction level on a Likert Scale with current geographical and environmental surroundings?

Writeup and dissemination, the final stages of the research process, should be obvious as they relate to life. Did you tell friends of your most recent grades? Have you used what you have learned to inform others? Are you now able to put letters after your name, proclaiming what you have accomplished?

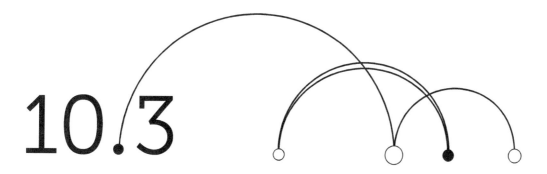

10.3

Why should you engage in research?

We have suggested you have been doing research all your life. Why quit now? If you do not buy that your life is a research study in progress (and we are only suggesting; no obligation to accept), we can nevertheless give you other reasons. Research does stimulate your brain, helps form new synaptic connections between neurons, even helps to build new neurons, and may help stave off old-age brain shrinkage. The creativity involved in coming up with the methodology and final design of a study is a good right-hemisphere wakeup that will help bring that portion of your brain into greater play.

We would also suggest that if you like puzzles, brainteasers, or word problems; finding solutions to why things do not work (jury-rigging your car or motorcycle); piddling around creating new stuff, you are a ripe candidate for research. We mentioned sleuthing way back in the book, in an analogy to Sherlock Holmes. Doing any of these things is sleuthing, so if you do any of them, you are a sleuther (new word). If you are a sleuther, you may well find research to your liking.

Are you altruistic? Do you think global warming may be an issue, that equality and respect for all are a fundamental right, and that blondes are no less-smart than others? Then you may be one of those people who wants to take a crack at advancing the margins of science, because all of these things are science-based. Science is a search for the truth. We may ever find a final truth with a capital "T," but it is such a search.

We will not rehash the hallowed ground argument here again. It stands on its own, and the names of those who have tread this ground are legion.

If we have waxed a bit poetic in this final chapter, do forgive us. Each of us truly believes that pursuit of knowledge in whatever form is valuable, can be romantic (you would be surprised how many couples have met over a dissertation), and is an endeavor of high importance.

10.4

Our hopes for you

First, we hope you have enjoyed the study of research methods and that we have made it more understandable. We hope we have sparked an interest, if not an understanding, of the importance of the process of research as a means to understanding your field of study and being an intelligent consumer of findings in your field—and other fields as well. In short, we hope this has been worthwhile for you.

Your authors have always enjoyed engaging in research. It is a challenge that many are not up to. The excitement is always there. Yes, research can be exciting! We have all formed lasting relationships with colleagues through our research, and those relationships have been a much-valued added benefit of research collaboration. And do not forget: Have research, will travel.

Best to you. Thanks for the opportunity to have been involved in the research of your life.

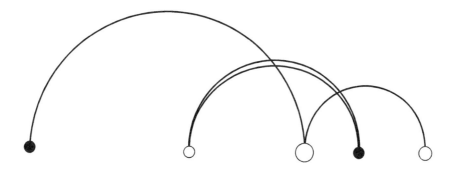

Appendix

Proportions of Area Under the Standard Normal Curve: The z-Tables

z	Area between Mean and z	Area beyond z in Tail	z	Area between Mean and z	Area beyond z in Tail
0.00	0.0000	0.5000	0.24	0.0948	0.4052
0.01	0.0040	0.4960	0.25	0.0987	0.4013
0.02	0.0080	0.4920	0.26	0.1026	0.3974
0.03	0.0120	0.4880	0.27	0.1064	0.3936
0.04	0.0160	0.4840	0.28	0.1103	0.3897
0.05	0.0199	0.4801	0.29	0.1141	0.3859
0.06	0.0239	0.4761	0.30	0.1179	0.3821
0.07	0.0279	0.4721	0.31	0.1217	0.3783
0.08	0.0319	0.4681	0.32	0.1255	0.3745
0.09	0.0359	0.4641	0.33	0.1293	0.3707
0.10	0.0398	0.4602	0.34	0.1331	0.3669
0.11	0.0438	0.4562	0.35	0.1368	0.3632
0.12	0.0478	0.4522	0.36	0.1406	0.3594
0.13	0.0517	0.4483	0.37	0.1443	0.3551
0.14	0.0557	0.4443	0.38	0.1480	0.3520
0.15	0.0596	0.4404	0.39	0.1517	0.3483
0.16	0.0636	0.4364	0.40	0.1554	0.3446
0.17	0.0675	0.4325	0.41	0.1591	0.3409
0.18	0.0714	0.4286	0.42	0.1628	0.3372
0.19	0.0753	0.4247	0.43	0.1664	0.3336
0.20	0.0793	0.4207	0.44	0.1700	0.3300
0.21	0.0832	0.4168	0.45	0.1736	0.3264
0.22	0.0871	0.4129	0.46	0.1772	0.3228
0.23	0.0901	0.4090	0.47	0.1808	0.3192

z	Area between Mean and z	Area beyond z in Tail
0.48	0.1844	0.3156
0.49	0.1879	0.3121
0.50	0.1915	0.3085
0.51	0.1950	0.3050
0.52	0.1985	0.3015
0.53	0.2019	0.2981
0.54	0.2054	0.2946
0.55	0.2088	0.2912
0.56	0.2123	0.2877
0.57	0.2157	0.2843
0.58	0.2190	0.2810
0.59	0.2224	0.2776
0.60	0.2257	0.2743
0.61	0.2291	0.2709
0.62	0.2324	0.2676
0.63	0.2357	0.2643
0.64	0.2389	0.2611
0.65	0.2422	0.2578
0.66	0.2454	0.2546
0.67	0.2486	0.2514
0.68	0.2517	0.2483
0.69	0.2549	0.2451
0.70	0.2580	0.2420
0.71	0.2611	0.2389
0.72	0.2642	0.2358
0.73	0.2673	0.2327
0.74	0.2704	0.2296
0.75	0.2734	0.2266
0.76	0.2764	0.2236
0.77	0.2794	0.2206
0.78	0.2823	0.2177
0.79	0.2852	0.2148
0.80	0.2881	0.2119
0.81	0.2910	0.2090
0.82	0.2939	0.2061
0.83	0.2967	0.2033
0.84	0.2995	0.2005
0.85	0.3023	0.1977
0.86	0.3051	0.1949
0.87	0.3078	0.1922
0.88	0.3106	0.1894
0.89	0.3133	0.1867
0.90	0.3159	0.1841
0.91	0.3186	0.1814

z	Area between Mean and z	Area beyond z in Tail
0.92	0.3212	0.1788
0.93	0.3238	0.1762
0.94	0.3264	0.1736
0.95	0.3289	0.1711
0.96	0.3315	0.1685
0.97	0.3340	0.1660
0.98	0.3365	0.1635
0.99	0.3389	0.1611
1.00	0.3413	0.1587
1.01	0.3438	0.1562
1.02	0.3461	0.1539
1.03	0.3485	0.1515
1.04	0.3508	0.1492
1.05	0.3531	0.1469
1.06	0.3554	0.1446
1.07	0.3577	0.1423
1.08	0.3599	0.1401
1.09	0.3621	0.1379
1.10	0.3643	0.1357
1.11	0.3665	0.1335
1.12	0.3686	0.1314
1.13	0.3708	0.1292
1.14	0.3729	0.1271
1.15	0.3749	0.1251
1.16	0.3770	0.1230
1.17	0.3790	0.1210
1.18	0.3810	0.1190
1.19	0.3830	0.1170
1.20	0.3849	0.1151
1.21	0.3869	0.1131
1.22	0.3888	0.1112
1.23	0.3907	0.1093
1.24	0.3925	0.1075
1.25	0.3944	0.1056
1.26	0.3962	0.1038
1.27	0.3980	0.1020
1.28	0.3997	0.1003
1.29	0.4015	0.0985
1.30	0.4032	0.0968
1.31	0.4049	0.0951
1.32	0.4066	0.0934
1.33	0.4082	0.0918
1.34	0.4099	0.0901
1.35	0.4115	0.0885

z	Area between Mean and z	Area beyond z in Tail	z	Area between Mean and z	Area beyond z in Tail
1.36	0.4131	0.0869	1.80	0.4641	0.0359
1.37	0.4147	0.0853	1.81	0.4649	0.0351
1.38	0.4162	0.0838	1.82	0.4656	0.0344
1.39	0.4177	0.0823	1.83	0.4664	0.0336
1.40	0.4192	0.0808	1.84	0.4671	0.0329
1.41	0.4207	0.0793	1.85	0.4678	0.0322
1.42	0.4222	0.0778	1.86	0.4686	0.0314
1.43	0.4236	0.0764	1.87	0.4693	0.0307
1.44	0.4251	0.0749	1.88	0.4699	0.0301
1.45	0.4265	0.0735	1.89	0.4706	0.0294
1.46	0.4279	0.0721	1.90	0.4713	0.0287
1.47	0.4292	0.0708	1.91	0.4719	0.0281
1.48	0.4306	0.0694	1.92	0.4726	0.0274
1.49	0.4319	0.0681	1.93	0.4732	0.0268
1.50	0.4332	0.0668	1.94	0.4738	0.0262
1.51	0.4345	0.0655	1.95	0.4744	0.0256
1.52	0.4357	0.0643	1.96	0.4750	0.0250
1.53	0.4370	0.0630	1.97	0.4756	0.0244
1.54	0.4382	0.0618	1.98	0.4761	0.0239
1.55	0.4394	0.0606	1.99	0.4767	0.0233
1.56	0.4406	0.0594	2.00	0.4772	0.0228
1.57	0.4418	0.0582	2.01	0.4778	0.0222
1.58	0.4429	0.0571	2.02	0.4783	0.0217
1.59	0.4441	0.0559	2.03	0.4788	0.0212
1.60	0.442	0.055	2.04	0.4793	0.0207
1.61	0.4463	0.0537	2.05	0.4798	0.0202
1.62	0.4474	0.0526	2.06	0.4803	0.0197
1.63	0.4484	0.0516	2.07	0.4808	0.0192
1.64	0.4495	0.0505	2.08	0.4812	0.0188
1.65	0.4505	0.0495	2.09	0.4817	0.0183
1.66	0.4515	0.0485	2.10	0.4821	0.0179
1.67	0.4525	0.0475	2.11	0.4826	0.0174
1.68	0.4535	0.0465	2.12	0.4830	0.0170
1.69	0.4545	0.0455	2.13	0.4834	0.0166
1.70	0.4554	0.0446	2.14	0.4838	0.0162
1.71	0.4564	0.0436	2.15	0.4842	0.0158
1.72	0.4573	0.0427	2.16	0.4846	0.0154
1.73	0.4582	0.0418	2.17	0.4850	0.0150
1.74	0.4591	0.0409	2.18	0.4854	0.0146
1.75	0.4599	0.0401	2.19	0.4857	0.0143
1.76	0.4608	0.0392	2.20	0.4861	0.0139
1.77	0.4616	0.0384	2.21	0.4864	0.0136
1.78	0.4625	0.0375	2.22	0.4868	0.0132
1.79	0.4633	0.0367	2.23	0.4871	0.0129

z	Area between Mean and z	Area beyond z in Tail
2.24	0.4875	0.0125
2.25	0.4878	0.0122
2.26	0.4881	0.0119
2.27	0.4884	0.0116
2.28	0.4887	0.0113
2.29	0.4890	0.011
2.30	0.4893	0.0107
2.31	0.4896	0.0104
2.32	0.4898	0.0102
2.33	0.4901	0.0099
2.34	0.4904	0.0096
2.35	0.4906	0.0094
2.36	0.4909	0.0091
2.37	0.4911	0.0089
2.38	0.4913	0.0087
2.39	0.4916	0.0084
2.40	0.4918	0.0082
2.41	0.4920	0.0080
2.42	0.4922	0.0078
2.43	0.4925	0.0075
2.44	0.4927	0.0073
2.45	0.4929	0.0071
2.46	0.4931	0.0069
2.47	0.4932	0.0068
2.48	0.4934	0.0066
2.49	0.4936	0.0064
2.50	0.4938	0.0062
2.51	0.4940	0.0060
2.52	0.4941	0.0059
2.53	0.4943	0.0057
2.54	0.4945	0.0055
2.55	0.4946	0.0054
2.56	0.4948	0.0052
2.57	0.4949	0.0051
2.58	0.4951	0.0049
2.59	0.4952	0.0048
2.60	0.4953	0.0047
2.61	0.4955	0.0045
2.62	0.4956	0.0044

z	Area between Mean and z	Area beyond z in Tail
2.63	0.4951	0.0043
2.64	0.4959	0.0041
2.65	0.4960	0.0040
2.66	0.4961	0.0039
2.67	0.4962	0.0038
2.68	0.4963	0.0037
2.69	0.4964	0.0036
2.70	0.4965	0.0035
2.71	0.4966	0.0034
2.72	0.4967	0.0033
2.73	0.4968	0.0032
2.74	0.4969	0.0031
2.75	0.4970	0.0030
2.76	0.4971	0.0029
2.77	0.4972	0.0028
2.78	0.4973	0.0027
2.79	0.4974	0.0026
2.80	0.4974	0.0026
2.81	0.4975	0.0025
2.82	0.4976	0.0024
2.83	0.4977	0.0023
2.84	0.4977	0.0023
2.85	0.4978	0.0022
2.86	0.4979	0.0021
2.87	0.4979	0.0021
2.88	0.4980	0.0020
2.89	0.4981	0.0019
2.90	0.4981	0.0019
2.91	0.4982	0.0018
2.92	0.4982	0.0018
2.93	0.4983	0.0017
2.94	0.4984	0.0016
2.95	0.4984	0.0016
2.96	0.4985	0.0015
2.97	0.4985	0.0015
2.98	0.4986	0.0014
2.99	0.4986	0.0014
3.00	0.4987	0.0013

Pearson Correlation Table

	One Tail		Two Tails				One Tail		Two Tails	
	0.05	**0.01**	**0.05**	**0.01**			**0.05**	**0.01**	**0.05**	**0.01**
$df = N - 2$	0.9877	0.9995	0.9969	0.9999		26	0.3172	0.4372	0.3739	0.4785
2	0.9000	0.9800	0.9500	0.9900		27	0.3115	0.4297	0.3673	0.4705
3	0.8054	0.9343	0.8783	0.9587		28	0.3061	0.4226	0.3610	0.4629
4	0.7293	0.8822	0.8114	0.9172		29	0.3009	0.4158	0.3550	0.4556
5	0.6694	0.8329	0.7545	0.8745		30	0.2960	0.4093	0.3494	0.4487
6	0.6215	0.7887	0.7067	0.8343		31	0.2913	0.4032	0.3440	0.4421
7	0.5822	0.7498	0.6664	0.7977		32	0.2869	0.3972	0.3388	0.4357
8	0.5494	0.7155	0.6319	0.7646		33	0.2826	0.3916	0.3338	0.4296
9	0.5214	0.6851	0.6021	0.7348		34	0.2785	0.3862	0.3291	0.4238
10	0.4973	0.6581	0.5760	0.7079		35	0.2746	0.3810	0.3246	0.4182
11	0.4762	0.6339	0.5529	0.6835		36	0.2709	0.3760	0.3202	0.4128
12	0.4575	0.6120	0.5324	0.6614		37	0.2673	0.3712	0.3160	0.4076
13	0.4409	0.5923	0.5140	0.6411		38	0.2638	0.3665	0.3120	0.4026
14	0.4259	0.5742	0.4973	0.6226		39	0.2605	0.3621	0.3081	0.3978
15	0.4124	0.5577	0.4821	0.6055		40	0.2573	0.3578	0.3044	0.3932
16	0.4000	0.5425	0.4683	0.5897		41	0.2542	0.3536	0.3008	0.3887
17	0.3887	0.5285	0.4555	0.5751		42	0.2512	0.3496	0.2973	0.3843
18	0.3783	0.5155	0.4438	0.5614		43	0.2483	0.3457	0.2940	0.3801
19	0.3687	0.5034	0.4329	0.5487		44	0.2455	0.3420	0.2907	0.3761
20	0.3598	0.4921	0.4227	0.5368		45	0.2429	0.3384	0.2876	0.3721
21	0.3515	0.4815	0.4132	0.5256		46	0.2403	0.3348	0.2845	0.3683
22	0.3438	0.4716	0.4044	0.5151		47	0.2377	0.3314	0.2816	0.3646
23	0.3365	0.4622	0.3961	0.5052		48	0.2353	0.3281	0.2787	0.3610
24	0.3297	0.4534	0.3882	0.4958		49	0.2329	0.3249	0.2759	0.3575
25	0.3233	0.4451	0.3809	0.4869		50	0.2306	0.3218	0.2732	0.3542

Spearman Rank Correlation Coefficient

	Two Tails		One Tail				Two Tails		One Tail	
	0.05	**0.01**	**0.05**	**0.01**			**0.05**	**0.01**	**0.05**	**0.01**
5	–	–	0.900	–		18	0.476	0.625	0.401	0.550
6	0.886	–	0.829	0.943		19	0.462	0.608	0.391	0.535
7	0.786	0.929	0.714	0.893		20	0.450	0.591	0.380	0.522
8	0.738	0.881	0.643	0.833		21	0.438	0.576	0.370	0.509
9	0.700	0.833	0.600	0.783		22	0.428	0.562	0.361	0.497
10	0.648	0.794	0.564	0.745		23	0.418	0.549	0.353	0.486
11	0.618	0.818	0.536	0.709		24	0.409	0.537	0.344	0.476
12	0.591	0.780	0.503	0.678		25	0.400	0.526	0.337	0.466
13	0.566	0.745	0.484	0.648		26	0.392	0.515	0.331	0.457
14	0.545	0.716	0.464	0.626		27	0.385	0.505	0.324	0.449
15	0.525	0.689	0.446	0.604		28	0.377	0.496	0.318	0.441
16	0.507	0.666	0.429	0.582		29	0.370	0.487	0.312	0.433
17	0.490	0.645	0.414	0.566		30	0.364	0.478	0.306	0.425

t Distribution Critical Values

	One Tail		Two Tails			One Tail		Two Tails	
	0.05	0.01	0.05	0.01		0.05	0.01	0.05	0.01
1	6.314	31.821	12.706	63.657	21	1.721	2.518	2.080	2.831
2	2.920	6.965	4.303	9.925	22	1.717	2.508	2.074	2.819
3	2.353	4.541	3.182	5.841	23	1.714	2.500	2.069	2.807
4	2.132	3.747	2.776	4.604	24	1.711	2.492	2.064	2.797
5	2.015	3.365	2.571	4.032	25	1.708	2.485	2.060	2.787
6	1.943	3.143	2.447	3.707	26	1.706	2.479	2.056	2.779
7	1.895	2.998	2.365	3.499	27	1.703	2.473	2.052	2.771
8	1.860	2.896	2.306	3.355	28	1.701	2.467	2.048	2.763
9	1.833	2.821	2.262	3.250	29	1.699	2.462	2.045	2.756
10	1.812	2.764	2.228	3.169	30	1.697	2.457	2.042	2.750
11	1.796	2.718	2.201	3.106	31	1.696	2.453	2.040	2.744
12	1.782	2.681	2.179	3.055	32	1.694	2.449	2.037	2.738
13	1.771	2.650	2.160	3.012	33	1.692	2.445	2.035	2.733
14	1.761	2.624	2.145	2.977	34	1.691	2.441	2.032	2.728
15	1.753	2.602	2.131	2.947	35	1.690	2.438	2.030	2.724
16	1.746	2.583	2.120	2.921	36	1.688	2.434	2.028	2.719
17	1.740	2.567	2.110	2.898	37	1.687	2.431	2.026	2.715
18	1.734	2.552	2.101	2.878	38	1.686	2.429	2.024	2.712
19	1.729	2.539	2.093	2.861	39	1.685	2.426	2.023	2.708
20	1.725	2.528	2.086	2.845	40	1.684	2.423	2.021	2.704

F Distribution alpha = .05

Degrees of Freedom within Groups	Degrees of Freedom between Groups						
	1	2	3	4	5	6	7
1	161.448	199.500	215.707	224.583	230.162	233.986	236.768
2	18.513	19.000	19.164	19.247	19.296	19.330	19.353
3	10.128	9.552	9.277	9.117	9.013	8.941	8.887
4	7.709	6.944	6.591	6.388	6.256	6.163	6.094
5	6.608	5.786	5.409	5.192	5.050	4.950	4.876
6	5.987	5.143	4.757	4.534	4.387	4.284	4.207
7	5.591	4.737	4.347	4.120	3.972	3.866	3.787
8	5.318	4.459	4.066	3.838	3.687	3.581	3.500
9	5.117	4.256	3.863	3.633	3.482	3.374	3.293
10	4.965	4.103	3.708	3.478	3.326	3.217	3.135
11	4.844	3.982	3.587	3.357	3.204	3.095	3.012
12	4.747	3.885	3.490	3.259	3.106	2.996	2.913
13	4.667	3.806	3.411	3.179	3.025	2.915	2.832

Degrees of Freedom within Groups	Degrees of Freedom between Groups						
	1	2	3	4	5	6	7
14	4.600	3.739	3.344	3.112	2.958	2.848	2.764
15	4.543	3.682	3.287	3.056	2.901	2.790	2.707
16	4.494	3.634	3.239	3.007	2.852	2.741	2.657
17	4.451	3.592	3.197	2.965	2.810	2.699	2.614
18	4.414	3.555	3.160	2.928	2.773	2.661	2.577
19	4.381	3.522	3.127	2.895	2.740	2.628	2.544
20	4.351	3.493	3.098	2.866	2.711	2.599	2.514
21	4.325	3.467	3.072	2.840	2.685	2.573	2.488
22	4.301	3.443	3.049	2.817	2.661	2.549	2.464
23	4.279	3.422	3.028	2.796	2.640	2.528	2.442
24	4.260	3.403	3.009	2.776	2.621	2.508	2.423
25	4.242	3.385	2.991	2.759	2.603	2.490	2.405
26	4.225	3.369	2.975	2.743	2.587	2.474	2.388
27	4.210	3.354	2.960	2.728	2.572	2.459	2.373
28	4.196	3.340	2.947	2.714	2.558	2.445	2.359
29	4.183	3.328	2.934	2.701	2.545	2.432	2.346
30	4.171	3.316	2.922	2.690	2.534	2.421	2.334
31	4.160	3.305	2.911	2.679	2.523	2.409	2.323
32	4.149	3.295	2.901	2.668	2.512	2.399	2.313
33	4.139	3.285	2.892	2.659	2.503	2.389	2.303
34	4.130	3.276	2.883	2.650	2.494	2.380	2.294
35	4.121	3.267	2.874	2.641	2.485	2.372	2.285
36	4.113	3.259	2.866	2.634	2.477	2.364	2.277
37	4.105	3.252	2.859	2.626	2.470	2.356	2.270
38	4.098	3.245	2.852	2.619	2.463	2.349	2.262
39	4.091	3.238	2.845	2.612	2.456	2.342	2.255
40	4.085	3.232	2.839	2.606	2.449	2.336	2.249
41	4.079	3.226	2.833	2.600	2.443	2.330	2.243
42	4.073	3.220	2.827	2.594	2.438	2.324	2.237
43	4.067	3.214	2.822	2.589	2.432	2.318	2.232
44	4.062	3.209	2.816	2.584	2.427	2.313	2.226
45	4.057	3.204	2.812	2.579	2.422	2.308	2.221
46	4.052	3.200	2.807	2.574	2.417	2.304	2.216
47	4.047	3.195	2.802	2.570	2.413	2.299	2.212
48	4.043	3.191	2.798	2.565	2.409	2.295	2.207
49	4.038	3.187	2.794	2.561	2.404	2.290	2.203
50	4.034	3.183	2.790	2.557	2.400	2.286	2.199

F Distribution alpha = .01

Degrees of Freedom within Groups	Degrees of Freedom between Groups						
	1	2	3	4	5	6	7
1	4052	5000	5403	5625	5764	5859	5928
2	98.49	99.00	99.17	99.25	99.30	99.33	99.34
3	34.12	30.82	29.46	28.71	28.24	27.91	27.67
4	21.20	18.00	16.69	15.98	15.52	15.21	14.98
5	16.26	13.27	12.06	11.39	10.97	10.67	10.45
6	13.74	10.92	9.78	9.15	8.75	8.47	8.26
7	12.25	9.55	8.45	7.85	7.46	7.19	7.00
8	11.26	8.65	7.59	7.01	6.63	6.37	6.19
9	10.56	8.02	6.99	6.42	6.06	5.80	5.62
10	10.04	7.56	6.55	5.99	5.64	5.39	5.21
11	9.65	7.20	6.22	5.67	5.32	5.07	4.88
12	9.33	6.93	5.95	5.41	5.06	4.82	4.65
13	9.07	6.70	5.74	5.20	4.86	4.62	4.44
14	8.86	6.51	5.56	5.03	4.69	4.46	4.28
15	8.68	6.36	5.42	4.89	4.56	4.32	4.14
16	8.53	6.23	5.29	4.77	4.44	4.20	4.03
17	8.40	6.11	5.18	4.67	4.34	4.10	3.93
18	8.28	6.01	5.09	4.58	4.25	4.01	3.85
19	8.18	5.93	5.01	4.50	4.17	3.94	3.77
20	8.10	5.85	4.94	4.43	4.10	3.87	3.71
21	8.02	5.78	4.87	4.37	4.04	3.81	3.65
22	7.94	5.72	4.82	4.31	3.99	3.76	3.59
23	7.88	5.66	4.76	4.26	3.94	3.71	3.54
24	7.82	5.61	4.72	4.22	3.90	3.67	3.50
25	7.77	5.57	4.68	4.18	3.86	3.63	3.46
26	7.72	5.53	4.64	4.14	3.82	3.59	3.42
27	7.68	5.49	4.60	4.11	3.79	3.56	3.39
28	7.64	5.45	4.57	4.07	3.76	3.53	3.36
29	7.60	5.42	4.54	4.04	3.73	3.50	3.33
30	7.56	5.39	4.51	4.02	3.70	3.47	3.30
31	7.53	5.36	4.48	3.99	3.67	3.45	3.28
32	7.50	5.34	4.46	3.97	3.65	3.43	3.26
33	7.47	5.31	4.44	3.95	3.63	3.41	3.24
34	7.44	5.29	4.42	3.93	3.61	3.39	3.22
35	7.42	5.27	4.40	3.91	3.59	3.37	3.20
36	7.40	5.25	4.38	3.89	3.57	3.35	3.18
37	7.37	5.23	4.36	3.87	3.56	3.33	3.17
38	7.35	5.21	4.34	3.86	3.54	3.32	3.15
39	7.33	5.19	4.33	3.84	3.53	3.30	3.14
40	7.31	5.18	4.31	3.83	3.51	3.29	3.12

Degrees of Freedom within Groups	Degrees of Freedom between Groups						
	1	2	3	4	5	6	7
41	7.30	5.16	4.30	3.82	3.50	3.28	3.11
42	7.28	5.15	4.29	3.80	3.49	3.27	3.10
43	7.26	5.14	4.27	3.79	3.48	3.25	3.09
44	7.25	5.12	4.26	3.78	3.47	3.24	3.08
45	7.23	5.11	4.25	3.77	3.45	3.23	3.07
46	7.22	5.10	4.24	3.76	3.44	3.22	3.06
47	7.21	5.09	4.23	3.75	3.43	3.21	3.05
48	7.19	5.08	4.22	3.74	3.43	3.20	3.04
49	7.18	5.07	4.21	3.73	3.42	3.20	3.03
50	7.17	5.06	4.20	3.72	3.41	3.19	3.02

Critical Values of Studentized Range Distribution (q_k) for Familywise alpha = .05

Degrees of Freedom within Groups	Number of Groups (a.k.a. Treatments)							
	3	4	5	6	7	8	9	10
1	26.976	32.819	37.081	40.407	43.118	45.397	47.356	49.070
2	8.331	9.798	10.881	11.734	12.434	13.027	13.538	13.987
3	5.910	6.825	7.502	8.037	8.478	8.852	9.177	9.462
4	5.040	5.757	6.287	6.706	7.053	7.347	7.602	7.826
5	4.602	5.218	5.673	6.033	6.330	6.582	6.801	6.995
6	4.339	4.896	5.305	5.629	5.895	6.122	6.319	6.493
7	4.165	4.681	5.060	5.359	5.606	5.815	5.997	6.158
8	4.041	4.529	4.886	5.167	5.399	5.596	5.767	5.918
9	3.948	4.415	4.755	5.024	5.244	5.432	5.595	5.738
10	3.877	4.327	4.654	4.912	5.124	5.304	5.460	5.598
11	3.820	4.256	4.574	4.823	5.028	5.202	5.353	5.486
12	3.773	4.199	4.508	4.748	4.947	5.116	5.262	5.395
13	3.734	4.151	4.453	4.690	4.884	5.049	5.192	5.318
14	3.701	4.111	4.407	4.639	4.829	4.990	5.130	5.253
15	3.673	4.076	4.367	4.595	4.782	4.940	5.077	5.198
16	3.649	4.046	4.333	4.557	4.741	4.896	5.031	5.150
17	3.628	4.020	4.303	4.524	4.705	4.858	4.991	5.108
18	3.609	3.997	4.276	4.494	4.673	4.824	4.955	5.071
19	3.593	3.977	4.253	4.468	4.645	4.794	4.924	5.037
20	3.578	3.958	4.232	4.445	4.620	4.768	4.895	5.008
21	3.565	3.942	4.213	4.424	4.597	4.743	4.870	4.981
22	3.553	3.927	4.196	4.405	4.577	4.722	4.847	4.957
23	3.542	3.914	4.180	4.388	4.558	4.702	4.826	4.935
24	3.532	3.901	4.166	4.373	4.541	4.684	4.807	4.915
25	3.523	3.890	4.153	4.358	4.526	4.667	4.789	4.897

Degrees of Freedom within Groups	Number of Groups (a.k.a. Treatments)							
	3	4	5	6	7	8	9	10
26	3.514	3.880	4.141	4.345	4.511	4.652	4.773	4.880
27	3.506	3.870	4.130	4.333	4.498	4.638	4.758	4.864
28	3.499	3.861	4.120	4.322	4.486	4.625	4.745	4.850
29	3.493	3.853	4.111	4.311	4.475	4.613	4.732	4.837
30	3.487	3.845	4.102	4.301	4.464	4.601	4.720	4.824
31	3.481	3.838	4.094	4.292	4.454	4.591	4.709	4.813
32	3.475	3.832	4.086	4.284	4.445	4.581	4.698	4.802
33	3.470	3.825	4.079	4.276	4.436	4.572	4.689	4.791
34	3.465	3.820	4.072	4.268	4.428	4.563	4.680	4.782
35	3.461	3.814	4.066	4.261	4.421	4.555	4.671	4.773
36	3.457	3.809	4.060	4.255	4.414	4.547	4.663	4.764
37	3.453	3.804	4.054	4.249	4.407	4.540	4.655	4.756
38	3.449	3.799	4.049	4.243	4.400	4.533	4.648	4.749
39	3.445	3.795	4.044	4.237	4.394	4.527	4.641	4.741
40	3.442	3.791	4.039	4.232	4.388	4.521	4.634	4.735
41	3.439	3.787	4.035	4.227	4.383	4.515	4.628	4.728
42	3.436	3.783	4.030	4.222	4.378	4.509	4.622	4.722
43	3.433	3.779	4.026	4.217	4.373	4.504	4.617	4.716
44	3.430	3.776	4.022	4.213	4.368	4.499	4.611	4.710
45	3.428	3.773	4.018	4.209	4.364	4.494	4.606	4.705
46	3.425	3.770	4.015	4.205	4.359	4.489	4.601	4.700
47	3.423	3.767	4.011	4.201	4.355	4.485	4.597	4.695
48	3.420	3.764	4.008	4.197	4.351	4.481	4.592	4.690
49	3.418	3.761	4.005	4.194	4.347	4.477	4.588	4.686
50	3.416	3.758	4.002	4.190	4.344	4.473	4.584	4.681

Formulas Used in Statistics

Formula Name	Formula
Frequency distribution	$rel\ f = \dfrac{f}{N}$
Percentile rank	$percentile = \dfrac{cf}{N}$
Interval size	$Interval\ Size = \dfrac{High\ score - Low\ score + 1}{Number\ of\ rows}$
Score at a given percentile	$Score = LRL + \dfrac{(target\ cf - cf\ below)}{f\ within}(Interval\ Size)$
Mean	$\bar{X} = \dfrac{\sum X}{N}$
Skewness	$sk = \dfrac{3(\bar{X} - M)}{s_x}$

Formula Name	Formula
Sample standard deviation conceptual formula	$S_x = \sqrt{\dfrac{\sum(X - \bar{X})^2}{N}}$
Sample standard deviation computational formula	$S_x = \sqrt{\dfrac{\sum X^2 - \dfrac{(\sum X)^2}{N}}{N}}$
Estimated population standard deviation	$s_x = \sqrt{\dfrac{\sum X^2 - \dfrac{(\sum X)^2}{N}}{N - 1}}$
Sample variance	$S_x^2 = \dfrac{\sum X^2 - \dfrac{(\sum X)^2}{N}}{N}$
Estimated population variance	$s^2 = \dfrac{\sum X^2 - \dfrac{(\sum X)^2}{N}}{N - 1}$
Z-score	$Z = \dfrac{X - \bar{X}}{s_x}$
Convert z-score to any score	$Score = (Z)(S_x) + \bar{X}$
Conceptual formula for Pearson r	$r_p = \dfrac{\sum Z_x Z_y}{N}$
Computational formula for Pearson r	$r_p = \dfrac{N\sum XY - (\sum X)(\sum Y)}{\sqrt{[N\sum X^2 - (\sum X)^2][N\sum Y^2 - (\sum Y)^2]}}$
Point-biserial correlation	$r_{pb} = \left(\dfrac{\bar{Y}_2 - \bar{Y}_1}{s_y}\right)(\sqrt{pq})$
Point-biserial significance	$t = \dfrac{r_{pb}\sqrt{n - 2}}{\sqrt{1 - r^2}}$
Spearman correlation	$r_s = 1 - \dfrac{6(\sum D^2)}{N(N^2 - 1)}$
Slope—regression	$b = \dfrac{N(\sum XY) - (\sum X)(\sum Y)}{N(\sum X^2) - (\sum X)^2}$
Intercept—regression	$a = \bar{Y} - b\bar{X}$
Regression	$Y' = a + bX$
Conceptual formula standard deviation of error	$Sy' = \sqrt{\dfrac{\sum(Y - Y')^2}{N}}$
Computational formula standard deviation of error	$S_{Y'} = S_Y \sqrt{1 - r^2}$
Standard error of the mean—single-sample z-test	$\sigma_{\bar{x}} = \dfrac{\sigma_x}{\sqrt{N}}$
Z-score for single-sample z-test	$Z = \dfrac{\bar{X} - \mu}{\sigma_{\bar{x}}}$
Standard error of the mean, estimated—single sample t-test	$S_{\bar{x}} = \dfrac{S_x}{\sqrt{N}}$

Formula Name	Formula
t-score for single-sample t-test	$$t = \frac{\bar{X} - \mu}{S_{\bar{x}}}$$
Standard error of the difference—independent-samples t-test	$$S_{\bar{x}_1 - \bar{x}_2} = \sqrt{\frac{(n_1 - 1)s_1^2 + (n_2 - 1)s_2^2}{(n_1 - 1) + (n_2 - 1)} \left(\frac{1}{n_1} + \frac{1}{n_2}\right)}$$
Independent-samples t-test conceptual formula	$$S_{\bar{x}_1 - \bar{x}_2} = \sqrt{\frac{(\sum \bar{X}_i - \bar{x}_j)^2}{N}}$$
t-score for Independent-samples t-test	$$t = \frac{\bar{X}_1 - \bar{X}_2}{S_{\bar{x}_1 - \bar{x}_2}}$$
Formulas for dependent t-test	
$s_D{}^2$—estimated variance of D	$$s_D^2 = \frac{\sum D^2 - \frac{(\sum D)^2}{N}}{N - 1}$$
$s_{\bar{D}}$—standard error of the mean difference	$$s_{\bar{D}} = \sqrt{s_D^2 \left(\frac{1}{N}\right)}$$
t-test for dependent t-test	$$t = \frac{\bar{D}}{s_{\bar{D}}}$$
Cohen's d or effect size for single-sample z-test	$$d = \frac{(\mu_1 - \mu)}{\sigma_x}$$
Cohen's d or effect size independent-samples t-test	$$d = \frac{(\mu_1 - \mu_2)}{\sigma_x}$$
σ—estimated population standard deviation for independent t effect size ($\tilde{S}_p = \sigma$)	$$\tilde{S}_p = \sqrt{\frac{(n_1 - 1)s_1^2 + (n_2 - 1)s_2^2}{n_1 + n_2 - 2}}$$
Cohen's d or effect size for dependent-sample t-test	$$d = \frac{(\mu_1 - \mu_2)}{\sigma_D}$$
σ_D for dependent-sample t-test effect size	$$\sigma_D = \sigma \sqrt{2(1 - r^2)}$$
Experiment wise error rate	$$\alpha_{ew} = (c)(\alpha_{pc})$$
F-ratio ANOVA	$$F = \frac{MS_{bn}}{MS_{wn}}$$
Within-group variance	$$MS_{wn} = \frac{\sum X_1^2 + \dots \sum X_k^2 - \left(\frac{(\sum X_1)^2}{n_1} + \dots \frac{(\sum X_k)^2}{n_k}\right)}{N - k}$$
Between-group variance	$$MS_{bn} = \frac{\frac{(\sum X_1)^2}{n_1} + \dots \frac{(\sum X_k)^2}{n_k} - \left[\frac{(\sum X_{tot})^2}{N}\right]}{k - 1}$$
Tukey (Honestly Significant Difference)	$$HSD = (q_k) \sqrt{\frac{MS_{wn}}{n}}$$

Formula Name	Formula
Fisher's protected t	$$t_{ij} = \frac{\bar{X}_i - \bar{X}_j}{\sqrt{MS_{wn}(\frac{1}{n_i} + \frac{1}{n_j})}}$$
One-way ANOVA effect size (omega squared)	$$\tilde{\omega}^2 = \frac{SS_{bn} - (k-1)(MS_{wn})}{SS_t + MS_{wn}}$$

Symbols Used in Statistics

Σ	Uppercase sigma—means to sum (add)
σ	Lowercase sigma—refers to population parameters; specifically, standard deviation
$\bar{X}$	Sample mean
μ	Population mean
N	Number of subjects in a sample
n	Subset of the subjects in a sample
k	Number of experimental groups
p	Probability
α	Level at which probability has been set (e.g., .05, .01, .001)
r	Correlation coefficient
t	A t-value (t-tests)
F	An F-value (Analysis of Variance: ANOVA)
ΣX	Add all the values for the variable X
ΣX^2	Square all the values for X, then sum those squared values
$(\Sigma X)^2$	Sum the value of X, then square that summed value
$(\Sigma X)(\Sigma Y)$	Sum the values of X and the values of Y, then multiply those two summed values
ΣXY	Multiply each individual X by its individual Y, then sum the results
$rel\ f$	Relative frequency
cf	Cumulative frequency
f	Frequency
LRL	Lower Real Limit
is	Interval size
sk	Coefficient of skewness
S_x	Sample standard deviation
s_x	Estimated population standard deviation
S^2	Sample variance
s^2	Estimated population variance

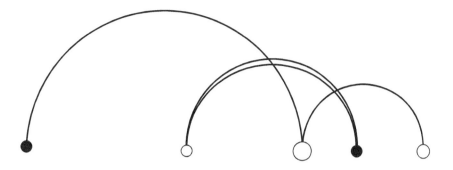

Glossary

Alternative hypothesis: used in traditional quantitative research to hypothesize that there is a relationship or significant difference between groups; opposite of the *null hypothesis*.

Annotated bibliography: list of references (books, journals, websites, periodicals etc.) that are relevant to a topic being researched.

ANOVA (Analysis of variance): statistical technique used to compare more than two random samples or groups at a time.

APA: American Psychological Association, normally referring to their publication manual, which is one of a number of guides for formatting and citations.

Archival data: data obtained from existing data sources, such as school or hospital records.

Blind studies (single or double): condition whereby aspects of the study are unknown in order to reduce any biases or predispositions of subjects or data collectors.

Causal factors: connections with a cause-effect relationship.

Citations: referenced sources used in a research project.

Confounding variables: variables, other than those under direct observation, that can influence the dependent variable and thus contribute to explain your findings.

Construct: comprehensive and overarching concept, particularly in psychology, to describe phenomena based on behavior, observations, or investigation.

Construct validity: degree to which a test measures what it claims to be measuring.

Content analysis: when a researcher examines the data for trends, patterns, themes that should exist.

Content validity: degree to which a measure represents all aspects of a given construct.

Continuous numbers: fractional or decimal numbers that can take any value within a range, such as a person's height or the length of a piece of wood.

Convenience sampling: sample selected from an easily accessible population, such as a survey at a mall where you are selected because you happened to be shopping there and walked near the person doing the survey.

Convergent validity: degree to which two measures of the same construct correlate to each other.

Correlation: mathematical expression of the degree of the relationship between two variables.

Cross-sectional study: observational technique where subjects are studied at a single point in time as opposed to over a span of time.

Dependent variable: outcome variable as measured on subjects.

Descriptive: statistics that describe the sample or population.

Design: statistical procedures used to answer a research question.

Dichotomous variable: variable with only two categories or levels (such as male or female).

Discrete numbers: whole numbers; that is, numbers without decimals or that are not fractions.

Distribution: spread of subjects across the measured variable, also known as frequency distribution.

Error: difference between exact mathematical value and observed value. All data theoretically have error.

Errors of commission: mistakes committed when the individual knows better regarding some procedure but chooses to deviate from accepted practices.

Errors of omission: mistakes that might also be termed "errors of ignorance" in that they are generally created when an individual collecting data does something wrong but is not aware of doing so.

Ethnocentrism: tendency for all of us to look at other cultural, ethnic, and racial groups through the lens of one's own culture.

Ethnographic methods: methods involving participant observations of individuals in their own environment using direct or indirect observations, interviews, focus groups, etc.

Experimental design: research design in which the researcher selects and assigns subjects randomly to treatment conditions.

Extraneous variable: a variable to exclude from a study (if possible) since it has the potential to explain any findings if not controlled.

Extrapolation: ability to apply a finding beyond a sample.

Face validity: perception that a test measures what it is intended to measure.

Frequency: number of times each score appears within a distribution.

Grounded theory analysis: method where trends, concepts, patterns, etc., will be revealed by the data as the data are reviewed and analyzed.

Hypothesis: statement or proposition about the characteristics or appearance of variables, or the relationship between variables, that acts as a working template for a particular research study.

Inferential techniques: statistical procedures that allow inferences to be made from the sample to the population.

Independent variable: researcher-controlled variable that is observed or manipulated in order to determine its effect on subjects.

Institutional review board (IRB): institutional committee that reviews and approves studies involving human subjects.

Interval data: true score data where you know the score a person made and you can tell the actual distance between individuals based on their respective scores, but the measure used to generate the score has no true zero.

Intervening variable: construct or phenomenon that provides a causal link between variables.

Hermeneutics: the method used to identify changes in organizations, expression, content, identity, voice, cultural underpinnings, etc. that might suggest a change in authors.

Kurtosis: deals with the height of a frequency distribution.

Likert Scale: measurement used to obtain the degree of agreement or disagreement by a subject.

Literature review: an examination of published works that outlines existing research about a subject.

Longitudinal study: long-term study in which the subjects are repeatedly observed over time with respect to the same variables

Measurement: way in which variables are defined and categorized.

Method: the technique used in a study, or the way of doing something.

Methodology: how tasks are done, not the doing. See *method*.

Mixed methods: research method that uses both quantitative and qualitative methods.

Model: representation of a system that purports to explain some phenomenon.

Moderating variable: variable that influences the relationship between other variables.

Multivariate techniques: statistical techniques that focuses on more than one variable or comparison of variables.

Negatively skewed: when bulk of subjects falls to the right side of the distribution.

Nominal: scale in which labels are assigned for identification but cannot be counted, such as male and female or categorical data where there may be more than two categories (e.g., Republican, Democrat, Independent).

Nonparametric: test of data where the data do not meet the assumptions of symmetry.

Null hypothesis: used in traditional quantitative research to predict that there is no relationship or significant difference between groups, opposite of the *alternative hypothesis*.

Operationalization: specifically defining the variables under investigation in a measurable way.

Ordinal: data (numbers) that indicate order only, but may not indicate what measurement was used to determine the order or the magnitude of the differences within the order.

Parametric techniques: statistics based on symmetrical distributions or distributions that come close to symmetry.

Parsimonious: logical tool used for theory development in which the fewest needed variables are used to provide the greatest explanatory power.

Peer review: system in which research is reviewed by experts in the field of study to ensure that the research was properly conducted.

Phenomenology: study of subjective experiences that may focus more on the underlying maintenance structures for these experiences.

Plagiarism: copying words, ideas, or findings from other researchers without providing credit.

Polytomies: variables with more than two levels or categories (such as Republican, Democrat, Independent).

Population: all members of a group; the larger group of individuals from which a sample can be selected.

Positively skewed: when bulk of subjects falls to the left side of the distribution.

Precise: how accurate a study is. Precision focuses the research question: cuts away competing chaff; increases viability of the investigation of the question by stating, without ambiguity, what one wants to know.

Predict: to make an assertion about what is expected to occur in a study.

Probability: chance that something will happen.

Quasi-experimental: design that cannot fully control for loss of internal or external validity due to risks that would go with the variable.

Random assignment: process of assigning subjects to treatment conditions in such a manner that each participant has an equal chance of being assigned to any particular group.

Random error: error in which unpredictable changes in the experiment occur by chance.

Random sampling: process of selecting a sample from the population, such that each member has an equal chance of being selected, in order to better represent the entire group as a whole and thus increase the generalizability of findings

Range: difference between the highest and lowest values.

Ratio: interval data with an absolute zero, such as height or weight.

Qualitative: research method used when data consist of words, not numbers; looks for the what, how, and why of some event or phenomenon.

Quantitative: research method involving numbers and data that can be manipulated mathematically and/or statistically.

Regions of rejections: area at tails of distribution curve where there is a low probability of an event occurring.

Reliability: consistency with which something is measured.

Replication: process to repeat an experimental condition to determine the variability associated with a phenomenon; could be carried out in an attempt to exactly match the initial study, or the methodology; could be duplicated but with a variation in the characteristics of the subjects.

Representative: extent to which a random sample mimics or looks like the larger group or population it came from.

Research: systematic way to study a particular subject; a process of adding to the existing knowledge in a particular field.

Research question: what you want to find out.

Sample: subset of subjects within the population who will be part of the study.

Scientific Method: basic guidelines that support all legitimate research. It starts with primarily three tenets, then goes on to describe the relevant characteristics of data and researchers.

Single-subject design: research design that uses only one subject.

Skew: measure of the symmetry of a distribution.

Snowball sampling: sampling based on the process of recommendation.

Statistics: study of methods for collecting, organizing, and analyzing numerical data.

Subjects: individuals in the sample; since no two subjects look exactly alike on all characteristics; differences in individual characteristics will increase the variance of the distribution.

Survey design: surveys or questionnaires used to collect data.

Systematic error: error caused by mistakes on a repeated basis that influence the accuracy of the measurement being made.

_t_-test: parametric test to test similarities or differences between samples or a sample and population when the population standard deviation is unknown.

Theory: statement of the inter-relationship of variables, behaviors, or events that purports to explain some phenomenon in detail.

Triangulation: analytical approach that brings two or more sources of data to bear on the research question.

Validity: whether a test is measuring what it is supposed to be measuring.

Variable: what is being measured or observed; phenomena that can take on different values, characteristics or categories.

Variance: mathematical indication of variability of subjects.

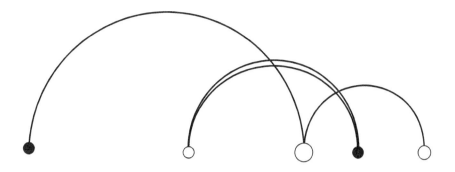

About the Authors

P. Michael Politano, PhD, MPS, ABPP

Dr. P. Michael (Mike) Politano is a professor in the Department of Psychology at the Citadel (Charleston, South Carolina). He holds an undergraduate degree from Duke University, a master's and PhD from Indiana University in school psychology, a postdoctorate from Indiana University and the Medical College of Virginia in clinical child psychology, and a master's in religious studies from Loyola, New Orleans. Dr. Politano has taught both undergraduate and graduate research methodology and statistics at various universities for more than 40 years. He also has supervised numerous master's theses and PhD dissertations.

Dr. Politano has published in numerous peer-reviewed journals and presented his work at multiple professional conferences worldwide. He is also the author and illustrator of the children's book *A Pig in a Tree* and the novel *Tag and Chubs*.

Robert O. Walton, PhD

Dr. Robert (Bob) O. Walton is an associate professor of business, executive director of international campus operations for Embry-Riddle Aeronautical University–Worldwide (ERAU), and managing director of Embry-Riddle Europe GmbH, a wholly owned subsidiary of ERAU. Dr. Walton received his undergraduate degree in geography from the University of North Carolina at Wilmington and holds a master of aeronautical science, master of business administration, master of logistics and supply chain management, and PhD in business administration.

Dr. Walton has published in numerous academic journals, authored or co-authored several books, and presented his research at multiple professional conferences worldwide. He was recognized as Faculty Member of the Year in 2010 and received the Outstanding Accomplishments in Research award from ERAU in 2012.

Donna L. Roberts, PhD

Dr. Donna Roberts is an associate professor in the College of Arts & Sciences at Embry-Riddle Aeronautical University-Worldwide (ERAU). She serves in the positions of interim department chair, discipline chair in Psychology & Sociology, and Undergraduate Research chair for the Department of Social Sciences & Economics, as well as on various academic committees within the university.

As a faculty member, Dr. Roberts has been involved in all aspects of the curriculum, from development to evaluation to delivery. She has previously served as dean of Academic Affairs for ERAU's International Region and as an officer of the Worldwide Faculty Senate.

Dr. Roberts holds master's degrees in counseling, human relations, adult and higher education, business administration, and aeronautical science, as well as a PhD in psychology.

Dr. Roberts's research interests include various areas of psychology and education, including personality, consumer psychology, aviation/aerospace psychology, educational psychology, and leadership psychology. She has published in various academic and professional journals and presented at numerous international conferences.

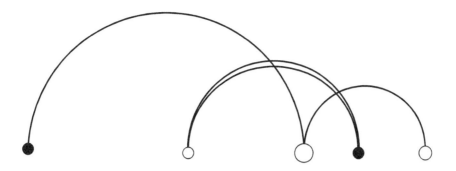

Index

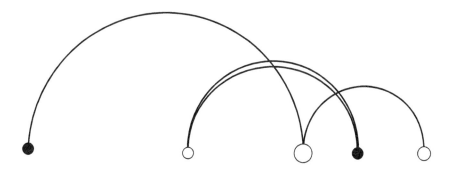

References

American Psychological Association. (2010). *Publication manual of the American Psychological Association* (6th ed.). Washington, DC: American Psychological Association.

Anderson, B. W. (1986). *Understanding the Old Testament.* Englewood Cliffs, NJ: Prentice-Hall, Inc.

Benedictow, O. L. (2005). The Black Death: The greatest catastrophe ever. *History Today, 55*(3), 42–48.

Boas, F. (1922). *Handbook of American Indian languages.* Washington, DC: Government Printing Office.

Bowlby, J. (1980). *Attachment and loss* (Vol. III). New York, NY: Basic Books.

Brainerd, C. J., & Reyna, V. F. (2012). Reliability of children's testimony in the era of developmental reversals. *Developmental Review, 32*(3), 224–267. doi: http://dx.doi.org?10.1016/j.dr.2012.06.008

Campbell, D. T., & Fiske, D. W. (1959). Convergent and discriminant validation by the multitrait-multimethod matrix. *Psychological Bulletin, 56*(2), 81–105.

Chiu, L. H. (1972). Cultural variations in emotions. *Psychological Bulletin, 112*(2), 179–204.

Cocks, C. (2001). *Doing the town.* Los Angeles, CA: University of California Press.

Cozby, P. C. (2009). *Methods in behavioral research* (10th ed.). New York, NY: McGraw-Hill.

Diener, E., Oishi, S., & Lucas, R. E. R. (2003). *Annual review of psychology, 54,* 4093–4425. doi: 10.1146/annurev.psych.54.101601.145056

Drennen, T. E., & Chapman, D. (1992). Greenhouse gases: Concentration on CO2, not methane from cows. *Choices, 7*(2), 31–32.

Edwards, R., & Holland, R. (2013). *What is qualitative interviewing?* New York, NY: Bloomsbury Publishing.

Eisenhart, M. (1991). Conceptual frameworks for research circa 1991: Ideas from a cultural anthropologist; implications for mathematics education researchers. Paper presented at the Proceedings of the Thirteenth Annual Meeting North American Chapter of the International Group for the Psychology of Mathematics Education, Blacksburg, VA.

Ekman, P. (1999). Basic emotions. In T. Dalgleish & M. Powers (Eds.), *Handbook of cognition and emotion*. Sussex, UK: John Wiley and Sons, Ltd.

Fetterman, D. M. (1998). *Ethnography step-by-step*. Thousand Oaks, CA: Sage.

Flynn, J. R. (1987). Massive IQ gains in 14 nations: What IQ tests really measure. *Psychological Bulletin, 101*(2), 171-191. doi: 10.1037/0033-2909.101.2.171

Fraser, R. A. R. (1995). How did Lincoln die? *American Heritage, 46*(1), 63–70.

Gardner, H. (1993). *Multiple intelligences: The theory in practice*. New York, NY: Basic Books.

Gay, P. (1988). *Freud: A life for our time*. New York, NY: W. W. Norton & Co.

George, D., & Mallery, P. (2012). *IBM SPSS statistics 19: Step by step*. Boston, MA: Pearson Education.

Gould, S. J. (1981). *The mismeasure of man*. New York, NY: Norton.

Graue, C. (2015). Qualitative data analysis. *International Journal of Sales, Retailing, and Marketing, 4*(9), 5–14.

Guralnik, D. B. (Ed.). (1980). *Webster's new world dictionary*. Cleveland, OH: William Collins Publishers.

Harris, R. J. (1985). *A primer of multivariate statistics*. New York, NY: Academic Press.

Heap, C. (2010). *Slumming: Sexual and racial encounters in American nightlife, 1885–1940*. Chicago, IL: University of Chicago Press.

Herubin, C. A. (1983). *Principles of surveying*. Englewood Cliffs, NJ: Prentice-Hall, Inc.

Jesson, J. K., Matheson, L., & Lacy, F. M. (2011). *Doing your literature review: Traditional and systematic techniques*. Los Angeles, CA: Sage Publications.

Ji, L. P., Nisbett, R. E., & Yonjie, S. (2001). Cultural change and prediction. *Psychological Science*, *12*(6), 450–456.

Johnson, L. T. (1999). *The writings of the New Testament: An interpretation*. Minneapolis, MN: Fortress Press.

Kerlinger, F. N. (1973). *Foundations of behavioral research* (2nd ed.). New York, NY: Holt, Rinehart and Winston.

Kerlinger, F. N. (1986). *Foundations of behavioral research* (3rd ed.). New York, NY: Holt, Rinehart, and Winston.

Lopez, B. (1986). *Arctic dreams*. New York, NY: Vintage Books.

Mabry, M. A. (1971). *The relationship between fluctuations in hemlines and stock market averages from 1921–1971*. Unpublished master's thesis. Knoxville, TN: University of Tennessee.

Matsumoto, D. (1994). *Cultural influences on research methods and statistics*. Pacific Grove, CA: Wadsworth Publishing.

Mayer, I. (2015). Qualitative research with a focus on qualitative data analysis. *International Journal of Sales, Retailing, and Marketing*, *4*(9), 53–67.

Mead, M. (1972). *Blackberry winter: My earlier years*. New York, NY: Kodansha International.

Mesquita, B., & Frijda, N. C. (1992). A cross-cultural comparison of cognitive styles in Chinese and American children. *Journal of Psychology*, *7*(4), 235–242.

National Commission for the Protection of Human Subjects of Biomedical and Behavioral Research (April 18, 1979). The Belmont report: Ethical principles and guidelines for the protection of human subjects of research. *Regulations and Ethical Guidelines*. Washington, DC: Department of Health, Education, and Welfare.

Nunnally, J. C. (1967). *Psychometric theory*. New York, NY: McGraw-Hill.

Owens, M., & Owens, D. (1992). *Cry of the Kalahari*. New York, NY: Mariner Books.

Preston, D. (2015). *Braddock's defeat: The battle of the Monongahela and the road to revolution*. New York, NY: Oxford University Press.

Rotter, J. B. (1990). Internal versus external control of reinforcement: A case history of a variable. *American Psychologist*, *45*(4), 489–493.

Schachter, S. (1963). Birth order, eminence, and higher education. *American Sociological Review*, *28*(5), 757–768.

Schwartz, S. (1986). *Classic studies in psychology*. Palo Alto, CA: Mayfield Publishing.

Segall, M. H., Campbell, D. T., & Herskovits, M. J. (1966). The *influence of culture on visual perception*. Indianapolis, IN: Bobbs-Merrill Co.

Sekaran, U., & Bougie, R. (2013). *Research methods for business: A skill building approach* (6th ed.). West Sussex, UK: John Wiley & Sons, Ltd.

Shneidman, E. (2004). *Autopsy of a suicidal mind*. New York, NY: Oxford University.

Simeonsson, R. I., & Boyles, E. K. (2001). An ecobiological approach in clinical assessment. In R. I. Simeonsson & E. K. Boyles (Eds.), *Psychological and developmental assessment: Children with disabilities and chronic conditions*. New York, NY: Guilford Press.

Tabachnick, B. G., & Fidel, L. (1989). *Using multivariate statistics* (2nd ed.). New York, NY: Harper and Row Publishers.

Tankard, J. W. (1984). *The statistical pioneers*. Cambridge, MA: Schenkman Publishing Co.

Tesch, R. (1990). *Qualitative research: Analysis types and software tools*. New York, NY: Falmer Press.

Thigpen, C. H., & Cleckley, H. (1954). A case of a multiple personality. *Journal of Abnormal Psychology, 49*(1), 135–151.

Thomas, A., Chess, B., & Birch, H. D. (1968). *Temperament and behavior disorders in children*. New York, NY: University of London Press.

Turner, M. B. (1967). *Psychology and the philosophy of science*. New York, NY: Appleton-Century-Crofts.

Walton, R. O. (2015). *Predicting financial distress in the all-cargo airline industry*. Saarbrücken, Germany: Lambert Publishing.

Walton, R. O., & Politano, P. M. (2014). Gender related perceptions and stress, anxiety, and depression on the flight deck. *Aviation Psychology and Applied Human Factors, 4*(2), 67-73. doi:10.1027/2192-0923/a000058

Walton, R. O., & Politano, P. M. (2016). Characteristics of general aviation accidents involving male and female pilots. *Aviation Psychology and Applied Human Factors, 6*(1), 39-44. doi: http://dx.doi.org/10.1027/2192-0923/a000085

Webb, E. J., Campbell, D. T., Schwartz, R. D., & Sechrest, L. (1965). *Unobtrusive measures: Nonreactive research in the social sciences*. Thousand Oaks, CA: Sage.

Weiner, P. P. (Ed.). (1966). *Charles S. Peirce: Selected writings*. New York, NY: Dover.